D0772711

Means of Escape from Fire

Means of Escape from Fire

M.J. Billington
Anthony Ferguson
and A.G. Copping

Blackwell
Science

Blackwell Science Ltd, a Blackwell Publishing Company
Editorial Offices:
Osney Mead, Oxford OX2 0EL, UK
Tel: +44 (0)1865 206206
Blackwell Science, Inc., 350 Main Street, Malden, MA 02148-5018, USA
Tel: +1 781 388 8250
Iowa State Press, a Blackwell Publishing Company, 2121 State Avenue, Ames, Iowa 50014-8300, USA
Tel: +1 515 292 0140
Blackwell Science Asia Pty, 54 University Street, Carlton, Victoria 3053, Australia
Tel: +61 (0)3 9347 0300
Blackwell Wissenschafts Verlag, Kurfürstendamm 57, 10707 Berlin, Germany
Tel: +49 (0)30 32 79 060

First published 2002 by Blackwell Science Ltd

Library of Congress
Cataloging-in-Publication Data is available

ISBN 0-632-03203-0

A catalogue record for this title is available from the British Library

Set in 11/13pt Bembo
by DP Photosetting, Aylesbury, Bucks.
Printed and bound in Great Britain by
MPG Books Ltd, Bodmin, Cornwall

For further information on
Blackwell Science, visit our website:
www.blackwell-science.com

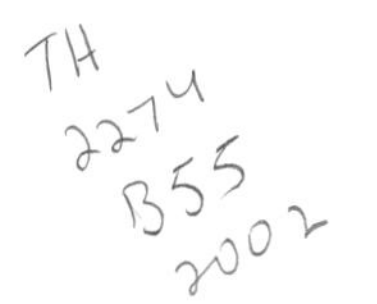

Contents

Preface

In order to protect lives in the event of a fire, it is absolutely essential that a building can be evacuated swiftly, safely and efficiently. The provision of an adequate means of escape is therefore fundamental to the design of new buildings, and to the alteration, change of use or extension of existing buildings. Additionally, with the passing of the Fire Precautions (Workplace) Regulations 1997 there is now a statutory duty on the employer in charge of a workplace to undertake an assessment of the fire risk. This includes consideration of the means of escape.

It is generally acknowledged that for maximum efficiency the means of escape should be based on the normal circulation routes in a building. Thus a building with a well-designed means of escape will function well and be convenient to use. Therefore it is essential that means of escape design principles are considered at the earliest stages of a project, when basic design decisions are being made relating to, for example, the positions of exits and entrances, corridor patterns and widths, staircase numbers and locations, and the need for lifts. Mistakes made at this stage can be very expensive to alter at later stages in the design, especially if the building control body or fire authority consider the means of escape to be unsatisfactory when the full design is submitted to them for comment and approval.

Inevitably, in view of the importance that it can have in the saving of lives, means of escape design and control has been the subject of a great deal of legislation scattered throughout a large number of statutes, regulations and supporting guidance documents. Many building uses are required to be licensed and/or registered in addition to the need for certification and Building Regulation compliance. It is therefore extremely difficult for designers, building owners and employers to be totally sure that they have complied with all relevant statutory requirements.

Therefore, in Chapters 1 to 3, we aim to identify the legislation which applies to any particular building use, and to show:

- what requirements need to be met
- when the requirements are applied to the building
- who is the controlling authority and what are their powers of enforcement
- which design standards are applied and what is their legal status.

Chapter 4 describes the general principles which apply to the design of means of escape in any building and will be of particular use to designers and those who are in charge of workplace fire safety.

Chapter 5 describes a 10 step approach to the design of means of escape which can be used for a range of residential and non-residential buildings, using the data contained in Approved Document B1 to the Building Regulations. This is the first time that the Approved Document guidance has been structured into a series of logical steps, which follow the design process. We have also attempted to simplify the degree to which cross-reference is necessary between different Approved Documents by providing a seamless integration of information from other parts of Approved Document B (such as the provision of fire doors) and from other Approved Documents (such as the data on minimum corridor widths from Approved Document M).

The mechanistic design solutions offered by Approved Document B may prove inadequate when dealing with large and complex buildings, such as hospitals and shopping complexes. Therefore, it may be more appropriate to use other sources of prescriptive design guidance, such as British Standards or specified codes of practice. Chapters 6 and 7 adopt this approach and give alternatives to the Approved Document B1 guidance for specific building uses.

There are many reasons why it can be difficult to design in full accordance with the prescriptive guidance outlined in Chapters 5 to 7. Chapter 8 describes how the safety objectives of means of escape legislation can be satisfied by modifying the basic principles, or by taking account of factors not included in the prescriptive guidance. This is intended to give readers a wider viewpoint and may prove useful when discussing various design options with fire legislation enforcement bodies.

In 1998, the British Standards Institute commissioned consultants to write the draft of a new British Standard (to be called BS 9999) on precautions against fire in buildings. The intention is that the new Standard will not only replace the BS 5588 series, but could fulfil the function of many other fire safety guidance documents that have traditionally been produced by government departments, and other bodies. Chapter 9 is based on the draft of Part 1: Means of Escape, as it went to public consultation in 2000. Although there may be significant changes before the Standard is finalised there are some interesting proposals in the draft of Part 1 that are worth describing now and there is no reason why some of the ideas should not be used in practice before the Standard is published.

For large and complex development schemes, even the prescriptive guidance given in the alternatives to Approved Document B1 may prove inadequate or inflexible. In such cases, the only course of action may be to propose a design based on a fire safety engineering strategy instead. Chapter 10 outlines the principles of this approach and will be of help to designers who are considering whether or not to adopt such a design strategy.

Most of this book is concerned with the design of means of escape. Obviously, even the highest design standards will be of little benefit if the building is not properly managed during its period of use. Additionally, there is now a statutory duty on an employer in charge of a workplace to undertake an assessment of the fire risk, and consideration of the means of escape will form part of this assessment. Chapter 11 addresses the issues of fire safety management in premises in use, and provides an example of a fire risk assessment strategy. This example will be of particular relevance to those in charge of fire safety management in an organisation since the procedure can be adapted to suit individual premises and can assist in determining compliance with the Fire Precautions (Workplace) Regulations.

This book is concerned primarily with the principles which govern the design and management of means of escape in case of fire. Therefore, with the exception of Chapter 5, which deals exclusively with Approved Document B1 and applies only to England and Wales, this book will be of use to designers, property professionals, building control bodies, fire officers and students in all parts of the United Kingdom. Additionally, since the statutory guidance in the Republic of Ireland makes extensive reference to British Standards much of the contents will also be relevant to the Republic.

The law is stated on the basis of cases reported and other material available to us on 30 November 2001.

Michael J Billington
Anthony Ferguson
Alexander Copping

Abbreviations

AD	Approved Document
ASET	available safe escape time
ATT	allowable travel time
BRAC	Building Regulations Advisory Committee
BS	British Standard
BSI	British Standards Institution
CCSs	Commissions for Care Standards
CIC	Construction Industry Council
DoE	Department of the Environment
DTLR	Department for Transport, Local Government and the Regions
FA	Factories Act
FPA	Fire Precautions Act
FSRCP	*Fire Safety in Residential Care Premises*
GLC	Greater London Council
HMOs	Houses in multiple occupation
HOFD	Home Office Fire and Emergency Planning Department
HSE	Health and Safety Executive
HSWA	Health and Safety at Work etc. Act
HTM	Health Technical Memorandum
LA	local authority
LDSA	London District Surveyors Association
MATT	maximum allowable travel time
NHBC	National House Building Council
OFSTED	Office for Standards in Education
OSRPA	Offices Shops and Railway Premises Act
PD	Published Document
PG	purpose group
QDR	qualitative design review
rcd	residual current device

Chapter 1
Means of Escape – The Background

1.1 Introduction

This year in England and Wales, fire brigades will be called to deal with around 100,000 fires in buildings. About 400 people will lose their lives and in the region of 14,000 people will suffer injury, most likely from burns or smoke inhalation. Yet, in 90% of these incidents the fire will not spread beyond the room of origin.

These statistics provide a stark illustration of the need for adequate means of escape in buildings and give a clue as to how it might be achieved – by giving people the ability to turn away from a fire and leave by means of a 'safe route', until a place of safety is reached.

This safe route will be most effective if it follows the normal circulation patterns of the building, since research has shown that when people react to a fire incident they tend to follow the routes with which they are most familiar. Therefore, a well-designed means of escape will usually provide an efficient circulation pattern for the building and vice versa.

Considerable savings in both design time and construction costs can be achieved for the client if means of escape design principles are considered at the earliest stages of a project when basic design decisions are being made relating to:

- the positions of exits and entrances;
- corridor patterns and widths;
- staircase numbers and locations; *and*
- the need for lifts.

Mistakes made at this stage can be very expensive to alter at later stages in the design, especially if the building control body or fire authority consider the means of escape to be unsatisfactory when the full design is submitted to them for comment and approval.

Whilst means of escape design is obviously of prime importance at the design stage of a new building, it is equally important for buildings in use. The law (see the Fire Precautions (Workplace) Regulations 1997) now places a statutory duty on an employer in charge of a workplace to undertake an assessment of the fire risk. This involves, among other things, an assessment of the adequacy of the means of escape. Therefore, employers also need a grasp of basic means of escape principles in order that they can fulfil their statutory obligations and avoid the penalties which exist under the Regulations, which can include a substantial fine and/or a prison sentence.

This book provides the grounding in means of escape design, which is vital for both designers and employers, and can be used to establish basic design principles for any building type.

1.2 Means of escape and the building life cycle

Means of escape is defined in British Standard 5588[1] as:

> 'structural means whereby a safe route or routes is or are provided for persons to travel from any point in a building to a place of safety.'

A place of safety is further defined as:

> 'a place in which persons are in no danger from fire.'

The concept of providing safe means of escape in buildings is well understood and a large number of controls exist to ensure that people may escape in the event of fire. This chapter aims to explain the background to the current systems of controls, show how they have come about and indicate where they might be headed in future. A detailed discussion of the legislation affecting means of escape may be found in Chapters 2 and 3.

A building passes through a number of stages during its life cycle. It starts as a design concept and then progresses through the construction, use and re-use phases before reaching obsolescence. It may then be refurbished or if this proves impractical it will be demolished and the site redeveloped. At each stage there is potential for it to be upgraded if this is thought to be desirable, feasible or essential in order to comply with changing standards, user requirements or increased public awareness of how a modern building should behave. This concept of 'change of state' is the most important principle in the formulation of means of escape legislation and is often a trigger for the assessment of the adequacy of existing means of escape arrangements in a building.

1.3 Means of escape and the new building

During the design and construction stages of a building, means of escape, in England and Wales, is subject to control under the provisions of paragraph B1 of Schedule 1 to the Building Regulations 2000[2]. The purpose of the Regulations is to secure adequate standards of health, safety, welfare and convenience for people in or about buildings or for others who may be affected by buildings. The Regulations may be administered by the building control department of the local authority or by a private approved inspector and the means of escape provisions of paragraph B1 apply to all buildings except prisons.

As will be seen in Chapter 2, the Regulations apply to new and extended buildings, to certain categories of altered buildings and to those which have their uses changed in certain narrowly defined ways. Therefore, there is no continuing control under the Regulations and once compliance has been demonstrated the local authority can only take action if they consider that further work is being carried out which contravenes the Regulations.

In fact, the use of Building Regulations to control means of escape is a relatively recent innovation and came about as a result of a report on fire safety prepared by the Holroyd Committee on the Fire Service[3] which was published in 1970. At that time it was becoming increasingly obvious that fire precautions law in general, and means of escape legislation in particular, were too fragmented due to the 'piecemeal' nature of their development. Much criticism was levelled at the lack of uniformity and the variety of enforcement and this led to demands for rationalisation and consolidation.

The main conclusion of the Holroyd Committee was that fire safety legislation should be consolidated into the following two main strands applying to:

- new and altered premises; *and*
- occupied premises.

The introduction of Part EE as an amendment to the 1972 Building Regulations (which came into force on 31 August 1973) went part way to satisfying the former recommendation although the regulations applied only to certain new flats, maisonettes, offices and shops. The 1976 Regulations consolidated the amendments to the 1972 edition without making any significant changes affecting means of escape and it was not until the issue of the 1985 Regulations that major revisions were made.

Following an extensive period of consultation, the publication of the Building Regulations 1985 marked a sea change in the way in which the mandatory requirements were presented. The former Regulations had been phrased in prescriptive terms and had been criticised as being too complex and difficult to understand. In particular, the system was perceived to be more cumbersome and bureaucratic than necessary. Additionally, the detailed form of the Regulations was blamed for inflexibility in use and was said to inhibit innovation and impose unnecessary costs.

In 1985 the Regulations were recast so that the mandatory requirements were expressed in functional terms. Methods of compliance were removed from the regulations and were provided in a series of non-mandatory 'Approved Documents', there being no obligation to follow the guidance contained in the Approved Documents if it could be demonstrated that an alternative approach would not result in contravention of the mandatory functional regulations. Additionally, in paragraph B1 of Schedule 1 to the 1985 Regulations, the opportunity was taken to extend the scope of control over means of escape to include dwellinghouses of three storeys and above in addition to flats, offices and shops. Curiously, although B1 was expressed in functional terms, it was thought necessary to make the guidance document mandatory. Apparently this situation arose because the Home Office doubted the ability of Building Control Officers to deal with means of escape on anything other than a prescriptive basis.

This situation continued until the 1991 Regulations came into force. The Building Regulations 1991 completed the process started in 1973 and extended control for means of escape to all buildings except prisons. They also brought paragraph B1 into line with the remainder of the Regulations and removed the mandatory nature of the guidance. This is now contained in Approved Document B (which also contains a great many other recommendations covering all aspects of fire safety in buildings).

The latest step in the process is contained in the Building Regulations 2000 whereby paragraph B1 has been extended to include means of warning in addition to means of escape.

Of course, the Building Regulations cover a great deal more than means of escape and a full coverage of the regulations and approved documents may be found in *The Building Regulations Explained and Illustrated* by Vincent Powell-Smith and M.J. Billington[4].

1.4 Means of escape and the building in use

Following on from the report of the Holroyd Committee, it was clear that new legislation would be required where occupied premises were concerned. Ultimately, this resulted in the introduction of the Fire Precautions Act in 1971[5] (FPA 1971) with the intention that the provisions of the Act would come into operation in stages and would eventually replace most of the existing legislation which applied to occupied premises. Initially hotels and boarding houses were the only building types designated under the Act, a multiple-fatality fire in a hotel having provided the impetus to get the Act on the statute books.

Unlike Building Regulation control the Fire Precautions Act requires that relevant premises be provided with a fire certificate. Occupiers who use premises without a certificate being in force (or without having applied for one) may leave themselves open to prosecution. This of course means that the fire authority must have the power of continuing control if they are to exercise their statutory function.

Interestingly, it was not until the introduction of the FPA 1971 that the Fire Service were given statutory responsibility for the enforcement of fire safety in buildings. Initially their role had been restricted to giving free advice on means of escape and other matters concerning fire prevention under section 1(1)(f) of the Fire Services Act 1947 (a role which still exists). The Factories Act 1971 (FA 1971) and the Offices Shops and Railway Premises Act 1963 (OSRPA 1963) did place a requirement on fire authorities to issue fire certificates for relevant premises; however the responsibility for enforcement was carried by the Factory Inspectorate.

Over the years the Fire Precautions Act has been substantially amended and extended. The Robens Report on Health and Safety at Work (1972)[6] recommended that the FPA 1971 should be extended to include all places of work and a step was made on the path to achieving this aim with the introduction of the Health and Safety at Work etc. Act 1974[7] (HSWA 1974). This Act added 'use as a place of work' to the uses listed in the FPA 1971. On 1 January 1977 the means of escape and fire precautions sections of the Factories Act 1971 and the Offices, Shops and Railway Premises Act 1963 were repealed (by virtue of the Fire Precautions (Factories, Offices, Shops and Railway Premises) Order 1976[8]) and became the responsibility of the fire authorities.

In the case of those factories, offices, shops and railway premises not requiring a fire certificate, certain fire safety requirements were imposed under the Fire Precautions (Non-Certificated Factory, Office, Shop and Railway Premises) Regulations 1976; however, both the 1976 Order and the Regulations were revoked and replaced in 1989[9].

A further rationalisation of fire safety law was necessitated when it became apparent that new legislation would be needed to implement the general fire safety provisions of the European Framework and Workplace Directives, which are not specifically dealt with by other legislation. This resulted in the enactment of the Fire Precautions (Workplace) Regulations 1997 (Fire (Workplace) Regulations)[10]. The Regulations require the provision of minimum fire safety standards in workplaces and impose duties on employers and on others in control of places of work. Initially, workplaces covered by other fire safety legislation (such as the Fire Precautions Act) were exempt from the requirements of the Fire (Workplace) Regulations. However, the Fire (Workplace) Regulations were amended in 1999[11] and now apply to any place of work where people are employed. The amendment was deemed necessary because the European Commission was concerned that the text of existing legislation applying to excepted workplaces in the UK did not implement the requirements of the Directives in full.

Since this results in dual application of the FPA 1971 and the Fire (Workplace) Regulations it has been necessary to include a number of safeguards in the Amendment Regulations so that, for example, the Fire Precautions Act cannot cause an employer to contravene any provision of the Fire (Workplace) Regulations. This legislation has far reaching implications for all workplaces and is discussed fully in Chapter 11 below.

A further recommendation of the Robens Committee affecting fire safety was that dangerous materials and processes should be covered by a comprehensive system of control administered by a new, unified safety inspectorate (the Health and Safety Executive). They also thought that in places constituting 'major hazards', responsibility for all fire matters, including means of escape and fire certification, should 'lie not with the fire authorities, but with the new inspectorate'. Thus, a splitting of control was suggested between *general fire precautions* administered by the fire authority on the one hand, and *process fire precautions* (where the concern is to minimise the risk of a fire starting and spreading because of the processes and activities carried out on the premises) on the other hand.

Again, the vehicle for this amendment was the HSWA 1974 which provided for control and enforcement by means of regulations. Using section 15 of the Act the Fire Certificates (Special Premises) Regulations 1976 were introduced and are the responsibility of the Health and Safety Executive with regard to the issue of fire certificates in premises constituting a major hazard.

A further amendment to the FPA 1971 was by virtue of the Fire Safety and Safety of Places of Sport Act 1987[12]. This enactment is much more extensive than its name would suggest and is not concerned exclusively with places of sport. Part 1 of the Act covers general fire precautions and its amendments to the 1971 Act are being introduced in stages. Some of the more significant changes include giving the fire authority powers to:

- exempt certain low risk premises from the requirement to have a fire certificate;
- charge for issuing and amending fire certificates;
- serve improvement notices on occupiers of exempt premises which do not have reasonable means of escape and fire-fighting facilities; *and*

- serve prohibition notices on the occupiers of designatable premises where there is a serious fire risk.

The effects of the Fire Safety and Safety of Places of Sport Act 1987 are considered in more detail in Chapter 3.

1.5 Criticism of the current systems of control

It is generally accepted that the Holroyd distinction between new buildings and occupied premises is a sound one on which to base the application of means of escape and fire safety legislation. This is because it is realised that there is more to fire safety than the physical characteristics of a building. Management of those fire safety characteristics is of equal importance in the day-to-day use of the building.

Nevertheless, the superficially simple distinction between new and occupied buildings has led to a number of difficulties in operation due to gaps and overlaps which exist in practice as follows:

- Although the Building Regulations purport to apply to all buildings there are a large number of exempted premises which could present a risk to the public, e.g. Crown buildings, some statutory undertakers' buildings, temporary structures, agricultural buildings, etc.
- Similarly, the FPA 1971 requires that a fire certificate be provided in only certain categories of designated buildings. This excludes assembly and recreation buildings, for example, on the grounds that they are covered by other legislation, such as the licensing of entertainments or of pubs, etc.
- The gaps identified have tended to be filled by the need to comply with a plethora of additional Acts and Statutory Provisions administered by different authorities at different stages in the life cycle of a building and to different guidance standards. Many building uses are required to be licensed and/or registered in addition to the need to be certificated and comply with the Building Regulations. This creates possible areas of overlap between the various legislative requirements, leads to confusion in the eyes of the public and the construction community, and can result in the omission of satisfactory control in some areas.
- The recent amendment to the Fire (Workplace) Regulations means that in most workplaces it is now possible for at least three pieces of fire safety legislation to apply to the same building, i.e. the Building Regulations during design and construction, and the Fire (Workplace) Regulations and the Fire Precautions Act during use. Although there are various safeguards in place to prevent conflicts of requirements, the current situation is bound to lead to confusion among designers and developers.

On this last point, in 1989 (and, in fact, before the Fire (Workplace) Regulations came into force), the Government decided to commission Bickerdike Allen Partners[13] to review the interrelationship between the Building Regulations and the fire certification requirements of the FPA 1971 and to examine the way they worked in practice. This review resulted in 1992 in the publication by the Department of the Environment, the Home Office and the Welsh Office of procedural guidance on Building Regulations and fire safety[14]. This guidance was intended to help designers, developers and occupiers of buildings to understand the steps involved in getting approval for fire safety aspects of building work and to explain the interaction between Building Regulations and other fire safety requirements.

The procedural guidance document sought to clarify the procedures for applicants by specifying a single point of contact at any particular time. This was to be the building control body (either the local authority or an approved inspector) for new and altered buildings, and the fire authority for premises after completion. Although the guidance document recommendations were not mandatory most authorities and building control bodies have adopted them.

The procedural guidance document was premised on the avoidance of dual application of the FPA 1971 and the Building Regulations by use of the statutory bar in section 13 of the FPA 1971. This prevents a fire authority requiring structural or other alterations to the means of escape if the Building Regulations also impose such requirements. However, the fire authority can require alterations:

- which are necessary to ensure that the premises comply with means of escape requirements of any regulations made under section 12 of the FPA 1971; *or*
- if they are not satisfied with the adequacy of the means of escape by virtue of matters not required to be shown in connection with provision of plans under the Building Regulations.

The coming into force of the Fire (Workplace) Regulations (under section 12 of the FPA 1971) has effectively removed the statutory bar for all workplaces covered by the Regulations. This means that in most circumstances it is possible for fire authorities to require structural and other alterations to means of escape even if they are deemed, by the building control body, to be in accordance with the Building Regulations.

Inevitably, the procedural guidance document has fallen into disuse and a revised version which caters for the changed circumstances is currently in preparation.

1.6 Means of escape – the way forward

It has been shown above that a system of fire precautions law has developed in response to individual needs (often in response to disaster) and this system now covers many different types of premises at all stages in their life cycles. Additionally, the scene has been further complicated by the need to comply with various European Community Directives on health and safety, culminating in the amendments to the Management of Health and Safety at Work Regulations 1992 brought about by the Fire Precautions (Workplace) Regulations 1997.

During the 1980s and 1990s a number of reviews and inquiries were set up to consider various aspects of fire safety law. Significant in the field of the fire safety review is the scrutiny report of the Interdepartmental Review Team, *Fire Safety Legislation and Enforcement*[15]. This was established on 17 January 1994 and its terms of reference included:

- a review of the operation and effectiveness of all legislation for which the Home Office, Department of the Environment and the Health and Safety Executive have policy responsibility in relation to fire safety;
- a review of the organisational arrangements of all the relevant enforcement bodies and agencies including recommendations for improvement; *and*
- addressing the practicability of bringing all policy responsibility for fire safety together under a single department.

The study was expected to identify any areas of overlap, duplication or lack of clarity between the responsibilities of enforcement bodies and was to make recommendations for dealing with them.

The final report was published in June 1994 and contained 61 conclusions and recommendations which addressed the criticisms outlined above. The report, which was widely circulated, elicited many replies during its consultation period. Although it was acknowledged as having gone some way towards simplifying procedures, many respondents felt that the report's insistence on the need for two enforcing authorities (building control bodies and fire authorities) would only perpetuate the existing weaknesses in the system.

A major factor in favour of the report was that it had been prepared by an interdepartmental review team representing the Department of Trade and Industry, the Department of the Environment and the Home Office. There was hope that a co-ordinated approach would help to break down some of the barriers which have existed between the various Government departments, since fire safety issues have always extended across departmental boundaries. Almost inevitably, the final report was something of a compromise, tending to expose protectionism within the Government departments rather than producing anything of real worth.

The intervention of the 1997 General Election, and a change of Government, resulted in a fresh appraisal of fire safety legislation. A consultative document (issued in November 1997) was entitled *Fire Safety Legislation for the Future*[16] and was published solely by the Home Office and Scottish Office. Its main proposals were:

- the establishment of a risk assessment driven, goal-based fire safety regime, imposing a universal duty of fire safety care;
- responsibility for ensuring satisfactory fire safety standards to be placed on the person responsible for the premises (similar to that for workplaces under the Fire (Workplace) Regulations);
- the proposed duty to extend to all those who may be on the premises (employees, members of the public and fire-fighters);
- the replacement of all existing fire safety provisions (consisting of more than 60 separate pieces of legislation) with a single Fire Safety Act;
- the regime to extend only to general fire precautions and not to process fire risks which would remain the responsibility of the Health and Safety Executive;
- enforcement of the Fire Safety Act to be the direct responsibility of fire authorities with built-in consultation requirements with other authorities;
- the replacement of the existing system of fire certification under the FPA 1971 with a form of fire authority validation of the responsible person's risk assessment;
- the provision of both simple non-technical, and detailed technical guidance, offering model solutions, but with the freedom to adopt alternatives if the fire safety goals are met.

The proposals had a mixed reception.

Since this consultative document was produced solely by the Home Office it is perhaps inevitable that the proposals sought to give greater powers to fire authorities (which came under the control of the Home Office at that time) thus increasing the importance of its own role in fire safety matters.

Whilst most consultees appeared to welcome the concept of imposing a general duty of fire safety care on employers, occupiers and owners of premises – all tied in to a new Fire Safety Act – the suggestion that all fire safety matters should be enforced by fire authorities caused much consternation, especially among building control bodies.

Following the 2001 General Election, there was a reorganisation of departmental responsibilities. It brought the Fire Service Inspectorate and the Building Regulations Division together, under the Department for Transport,

Local Government and the Regions (DTLR). There may therefore be a unified approach to fire safety in the future.

Thus, in the field of fire safety legislation there is a general movement towards simplification and a genuine desire to place the responsibility for fire safety matters in the hands of those who control the day-to-day use of buildings. It is to be hoped that the attainment of such goals will not be held back by the vested interests of the parties involved in the enforcement process, although this is less likely to happen now that all fire regulatory responsibilities are within one Government department.

1.7 References

1 BS 5588 : Fire precautions in the design, construction and use of buildings.
2 The Building Regulations 2000, (SI 2000/2531).
3 Report of the Departmental Committee on the Fire Service (Cmnd 4371, 1970).
4 Vincent Powell-Smith and M.J. Billington, *The Building Regulations Explained and Illustrated*, Blackwell Science, Oxford, 1999.
5 Fire Precautions Act 1971, c.40.
6 Health and Safety at Work (Cmnd 5034, 1972).
7 Health and Safety at Work etc. Act 1974, c.37.
8 See the Fire Precautions (Factories, Offices, Shops and Railway Premises) Order 1976 (SI 1976/2009).
9 See the Fire Precautions (Factories, Offices, Shops and Railway Premises) Order 1989 (SI 1989/76) and the Fire Precautions (Factories, Offices, Shops and Railway Premises) (Revocation) Regulations 1989 (SI 1989/78).
10 The Fire Precautions (Workplace) Regulations 1997 (SI 1997/1840).
11 The Fire Precautions (Workplace) (Amendment) Regulations 1999 (SI 1999/1877).
12 Fire Safety and Safety of Places of Sport Act 1987, c.27.
13 Review of the interrelationship between Building Regulations and other fire safety requirements, Bickerdike Allen and Partners, 1990.
14 *Building Regulation and Fire Safety – Procedural Guidance*, DoE, Home Office and Welsh Office, 1992.
15 *Fire Safety Legislation and Enforcement.* Report of the Interdepartmental Review Team, 1994.
16 *Fire Safety Legislation for the Future*, Home Office, 1997.

Chapter 2

New and Altered Buildings – the Statutory Requirements

2.1 Introduction

The purpose of this chapter is to set out in detail the principal legal controls over means of escape in new and altered buildings. It has already been shown that means of escape legislation tends to be applied to a building at various stages during its life cycle, when a 'change of state' takes place. Therefore if the relevant legislation is examined it is possible to determine:

- what requirements need to be met;
- when the requirements are applied to the building;
- who is the controlling authority and what are their powers of enforcement;
- which design standards are applied and what is their legal status.

This last point is dealt with in detail in the subsequent chapters of this book since often there are a number of ways of satisfying the legal requirements using various codes, standards and other guidance documents.

2.2 The Building Act 1984 and the Building Regulations 2000

2.2.1 Building Regulations – control over means of escape

The Building Act 1984[1] (as amended) contains the statutory framework for the building control system generally, covering the making, application, administration and enforcement of building regulations. Certain sections of the Act are highly relevant to means of escape and are discussed in more detail below.

The primary control over fire safety in general and means of escape in particular in new and altered buildings is through the medium of The Building

Regulations 2000[2] (as amended) which came into force on 1 January 2001. Building Regulations are made under powers contained in section 1 of the Building Act 1984 by the Secretary of State for Transport, Local Government and the Regions and may be made for the following broad purposes:

(1) securing the health, safety, welfare and convenience of persons in or about buildings and of others who may be connected with buildings;
(2) furthering the conservation of fuel and power; and
(3) preventing waste, undue consumption, misuse or contamination of water.

The Regulations so made, may cover the design and construction of buildings and the provision of services, fittings and equipment in or in connection with buildings.

When considering means of escape, we are concerned only with (1) above and Regulation 8 makes it clear that the power of the Secretary of State to make regulations about means of escape is limited to securing reasonable standards of health and safety for people who use buildings. The legal requirement for means of escape is contained in Schedule 1 to the 2000 Regulations, paragraph B1:

> 'The building shall be designed and constructed so that there are appropriate provisions for the early warning of fire, and appropriate means of escape in case of fire from the building to a place of safety outside the building capable of being safely and effectively used at all material times.'

It should be noted that despite the reference to the place of safety being *outside* the building it is sometimes desirable, such as in hospitals, for this to be within the building, but protected from the effects of fire. This is discussed more fully in subsequent chapters. Additionally, the requirement for means of escape does not apply to prisons provided under section 33 of the Prisons Act 1952[3]. (See also below for further exemptions.)

Therefore, it is not the purpose of the Regulations to protect property. This raises an interesting point since the current philosophy of fire safety law is to protect life. The view is that property protection is the concern of the individual and the insurance industry, though clearly, many life safety measures do add to property protection. In fact, occupiers or owners of buildings may find that having satisfied the legal requirements for fire safety and means of escape they are required to upgrade their premises still further in order to satisfy their insurers. (For guidance on property protection see *LPC Design guide for the fire protection of buildings*[4].)

2.2.2 *Building Regulations – form and content*

It is useful to regard the Regulations as being divided into the following main sections:

(1) the procedural requirements containing the administrative rules consisting of 24 numbered Regulations;
(2) the substantive requirements in Schedule 1 containing the technical provisions expressed in functional terms; *and*
(3) details of exempt buildings and work contained in Schedule 2.

The Schedule 1 functional requirements are grouped alphabetically in 13 parts and cover much more than just means of escape. However there are a number of parts which are particularly relevant to a consideration of fire safety:

- *Part A: Structure* covers loading, ground movement and disproportionate collapse.
- *Part B: Fire Safety* deals with means of warning and escape, internal and external fire spread, and access and facilities for the fire service.
- *Part F: Ventilation* covers means of ventilation and avoidance of condensation.
- *Part J: Combustion Appliances and Fuel Storage Systems* deals with air supply, discharge of products of combustion, protection of the building, provision of information, protection of liquid fuel storage systems from fire and protection against pollution.
- *Part K: Protection from Falling, Collision and Impact* covers stairs, ladders and ramps, protection from falling and vehicle barriers and loading ramps, protection from collision with open windows, etc. and protection against impact from and trapping by doors.
- *Part L: Conservation of Fuel and Power* requires reasonable provision to be made for the conservation of fuel and power.
- *Part M: Access and Facilities for Disabled People* deals with the access and use of the building, sanitary conveniences and audience or spectator seating.

2.2.3 *Building Regulations – practical guidance on compliance*

Since the Building Regulations are phrased in functional terms they merely state what is required without giving details of how the requirement may be satisfied. For example, paragraph K1 (stairs, ladders and ramps) of Part K to Schedule 1 states that:

> 'Stairs, ladders and ramps shall be so designed, constructed and installed as to be safe for people moving between different levels in or about the building.'

Clearly, people may have different opinions as to what constitutes a safe stair, and similar considerations may exist for all the other Parts of the Regulations. Therefore, a number of 'official' design guides have been produced for use with each Part of the Regulations and these may be used at the designer's discretion. Among the most common of these are British Standards produced by the British Standards Institution, and Approved Documents produced by the Department for Transport, Local Government and the Regions and the Welsh Office after consultation with the Building Regulations Advisory Committee (BRAC).

In fact, there are 14 Approved Documents corresponding to the 13 Parts of Schedule 1 to the Building Regulations, plus one other which deals with materials and workmanship and gives practical guidance on the requirements of Regulation 7. The Approved Documents are written in straightforward technical terms and include diagrams for ease of reference. They are updated at regular intervals after full consultation has taken place involving BRAC and interested parties from all areas of the construction industry.

The legal status and use of the Approved Documents is laid down in sections 6 and 7 of the Building Act 1984. Section 6 allows the Secretary of State or a body designated by him to approve and issue documents 'providing practical guidance with respect to the requirements of any provision of building regulations'. The legal effect of Approved Documents is contained in section 7. Their use is not obligatory and failure to comply with their requirements does not automatically involve any civil or criminal liability. However, if a designer or contractor decides against their use then they must bear the responsibility for satisfying the regulation requirements in some other way.

In the case of *Rickards* v. *Kerrier District Council* (1987)[5], it was held that if the local authority proved that the work was not in compliance with the Approved Documents, then it was up to the appellant to show that the requirements of the Regulations had still been met.

Clearly, the recommendations of Approved Document B (AD B) are of direct relevance to means of escape; however the approved documents which correspond to the other Parts of the Building Regulations listed above contain important recommendations which in some cases appear to contradict those of AD B. For example, the recommended width of an escape route from AD B may be shown to be less than that needed as access for disabled people in AD M. This problem of conflicting recommendations is common in Building Regulations and is solved, to a certain extent, by reference to Regulation 4(1)(b) where building work is required to be carried out so that – 'in complying with any such requirement [of Schedule 1] there is no failure to comply with any other such requirement'. In other words, it is necessary always to consider the most onerous case where a conflict exists.

2.3 Exempted buildings and work

2.3.1 Crown immunity

The Building Regulations do not apply to premises which are occupied by the Crown. The general position regarding Crown exemption was established in the case of *Gorton Local Board* v. *Prison Commissioners* (1887)[6] where it was held that 'a statute does not bind the Crown unless it so provides either expressly or by necessary implication'. In fact, there is provision in section 44 of the Building Act 1984 to apply the substantive requirements of the Regulations to Crown buildings but this has never been activated.

A definition of 'Crown property' may be found in section 87 of the 1984 Act and includes:

> 'any house, building or other premises being property belonging to Her Majesty in right of the Crown or of the Duchy of Lancaster, or belonging to a government department, or held in trust for Her Majesty for purposes of a government department.'

In recent years a number of premises have lost their Crown immunity, often in response to public opinion, the most notable of these being National Health Service buildings (see section 60 of the National Health and Community Care Act 1990).

The general exemption of Crown buildings can raise problems where, for example, one floor of a new, multi-storey office block is to be occupied by a government department (such as the Inland Revenue). Since the staircases form part of the means of escape it is possible for a section of a continuous staircase to be exempt from the Regulations! In practice, it is normal for government department building work to be designed and constructed in accordance with the Building Regulations. In some areas the plans and particulars may even be submitted to the local authority for comment, although it is more usual for these to be scrutinised by specialist companies such as Carillion Specialist Services Ltd[7] who will also carry out on-site inspections of the works in progress. Even so, such companies have no legal control over the work and cannot take enforcement action in the event of a breach of the Regulations.

Interestingly, Crown premises are not exempt from certification under the Fire Precautions Act 1971 (see Chapter 3). However, they are inspected not by the fire authority, but by the Crown Premises Inspection Group within the

Home Office Fire Service Inspectorate, a bureaucratic anomaly which has attracted much criticism. Unfortunately, the powers of entry to premises contained in the 1971 Act do not apply to premises occupied by the Crown.

2.3.2 Building Act exemptions

Taken together, the Building Act 1984 and the Building Regulations 2000 exempt certain uses of buildings and many categories of work from control. Unfortunately, the situation with regard to exempt building uses is often unclear. Section 4 of the 1984 Act (as amended) removes the following from Building Regulation control:

- Building work at maintained schools approved by the Secretary of State for Education and Employment. With the coming into force of the Education (Schools and Further and Higher Education)(Amendment)(England) Regulations 2001[8] on 1 April 2001, the Secretary of State's approval is no longer required and all schools and further and higher education buildings in England are now subject to the provisions of the Building Regulations 2000.
- Building work to premises belonging to statutory undertakers, the UK Atomic Energy Authority, or the Civil Aviation Authority. The building must be one which is held and used for the purpose of the undertaking (such as a sewage pumping station), therefore buildings such as houses are not exempt, and neither are showrooms or offices unless they are part of a railway station or (in the case of the Civil Aviation Authority) are on an aerodrome which is owned by the Authority. Such showrooms and offices would still have to be used for the purposes of that undertaking to be exempt.

Regarding statutory undertakers, the fact that many former public bodies are now in private hands has meant that both gas and electricity suppliers are not now regarded as statutory undertakers under the Act. Furthermore, by virtue of the Postal Services Act 2000[9] Post Offices (including other universal providers of postal services) are no longer regarded as statutory undertakers and thus, are required to submit details regarding controlled work under the Building Regulations to a building control body. The National Rivers Authority and water and sewerage undertakers are still deemed to be statutory undertakers for the purposes of the Act.

Under section 5 of the Building Act 1984, Building Regulations may exempt certain public bodies from the procedural requirements of the regulations. These include:

- local authorities;
- county councils; *and*
- any other public body acting under statute for public purposes and not for its own profit.

To date the Metropolitan Police Authority is the only organisation to be designated as an exempt body for the purposes of the Act (see Building Regulation 10). The practical effect of this is that the Metropolitan Police Authority is still required to comply with the substantive requirements of the Regulations regarding the design and construction of buildings and the provision of services, fittings and equipment; however it does not need to submit particulars for approval by a building control body and it is not subject to local authority enforcement procedures.

It should be noted that local authority buildings are exempt from either the procedural or substantive requirements of the Building Regulations.

2.3.3 Building Regulation exemptions

By virtue of Regulation 9 certain buildings and extensions are granted exemption from control and are listed in Schedule 2. Additionally, if work is carried out on an exempt building, that work is itself exempt if on completion the building remains as described in Schedule 2. The exempt buildings and work fall into seven classes as follows:

Class I – Buildings controlled under other legislation

- Buildings subject to the Explosives Act 1875[10] and 1923[11].
- Buildings (but not associated dwellings, offices or canteens) on a site licensed under the Nuclear Installations Act 1965[12].
- Buildings scheduled under section 1 of the Ancient Monuments and Archaeological Areas Act 1979[13].

Class II – Buildings not frequented by people

- Detached buildings into which people do not normally go or only enter intermittently in order to inspect or maintain fixed plant or machinery.

These buildings are only exempt if they are at least one and a half times their own height from the site boundary or any other building which is frequented by people.

Class III – Greenhouses and agricultural buildings

- A building used as a greenhouse or agricultural building (including the keeping of animals).

However, buildings which are used principally for retailing, packing or exhibiting are not exempted. Agriculture is defined to include horticulture, fruit growing, the growing of plants for seed and fish farming. It should be noted that in order to remain exempt, an agricultural building must not be used as a dwelling and must be sited at least one and a half times its height from any building containing sleeping accommodation *and*, must be provided with a fire exit within 30 m of any point in the building.

Class IV – Temporary buildings

- Any building which is intended to be retained for no more than 28 days.

Class V – Ancillary buildings

- Buildings used in connection with the sale of buildings or building plots which are erected on the site.
- Site buildings on civil engineering and construction sites provided that they contain no sleeping accommodation.
- Any site building used in connection with mining or quarrying provided that it does not contain a dwelling, office or showroom.

Class VI – Small detached buildings

- Detached single-storey buildings without sleeping accommodation and floor area not exceeding 30 m^2.

For the exemption to apply the building must be constructed substantially of non-combustible material *or* be situated at least 1 m from the boundary of its curtilage.

- Detached buildings with floor area not greater than 30 m^2 used only to shelter people from the effects of nuclear, chemical or conventional weapons. Since these shelters are usually placed underground, the building must be kept at least the depth of the excavation plus 1 m from any other building or structure.
- Detached buildings without sleeping accommodation and floor area not exceeding 15 m^2.

Class VII – Extensions

- Ground level extensions consisting of conservatories, porches, covered yards or ways, and carports open on at least two sides with floor area not greater than 30 m^2.

However, a substantially glazed porch or conservatory must comply with Part N (Glazing – Safety in relation to impact, opening and cleaning) of Schedule 1 to the Building Regulations in order to be exempt.

2.4 The application of Building Regulations to projects

Assuming that a particular scheme is not exempt from the provisions of the Building Regulations by virtue of the above, it is still necessary to determine exactly how that scheme needs to comply with the procedural and substantive requirements since the degree of compliance will depend on the actual work being carried out.

The Building Regulations apply to 'building work' which is defined in Regulation 3 as:

- the erection or extension of a building;
- the provision or extension of a controlled service or fitting;
- the material alteration of a building or of a controlled service or fitting;
- work in consequence of a material change of use;
- the insertion of insulation into a cavity wall; *and*
- underpinning of a building.

2.4.1 Erection or extension of a building

No attempt is made in the Regulations to define erection or extension; however, section 123 of the Building Act 1984 indicates that this can include the reconstruction of a building or the roofing over of a space between walls or buildings. Additionally, the conversion of a movable object into a building can also be regarded as the erection of a building.

2.4.2 Controlled services or fittings

This includes sanitary conveniences and bathroom fittings, unvented hot water storage systems, drainage and waste disposal systems, heat producing appliances burning solid fuel, oil and gas, and certain services and fittings covered by Part L (Conservation of fuel and power). Although electrical installations are not explicitly covered in the Regulations, Approved Document B contains recommendations for electrical emergency lighting and warning systems, etc. on escape routes, which could be controlled under Part B1 (Means of escape).

2.4.3 Material alterations

If an alteration is carried out to an existing building or to a controlled service or fitting which at *any* stage results in:

- non-compliance with a relevant requirement where previously there was compliance; *or*
- a non-complying relevant requirement being made worse

then that alteration is considered to be material for the purposes of the Building Regulations and it constitutes building work to which the Regulations apply.

The relevant requirements include only the following sections of Schedule 1:

- Part A (Structure);
- in Part B (Fire Safety) the following paragraphs only: B1 (means of warning and escape), B3 (internal fire spread – structure), B4 (external fire spread) and B5 (access and facilities for the fire service);
- Part M (access and facilities for disabled people).

Therefore, the applicant or his advisor must judge whether or not the work constitutes a material alteration. In practice, much confusion has arisen over the phrase 'at any stage' since, for example, mere redecoration of premises could be viewed as a material alteration if the contractor temporarily blocked the means of escape with trestles or ladders in order to access the walls and ceilings. In order to be sure, it is best always to consult a building control body (local authority or approved inspector) if it is considered that the work would affect the sections mentioned above. In theory, if a submission is made to a local authority they should return the submission and any fee paid if they do not consider that the proposed alterations are material.

2.4.4 *Material change of use*

Not all changes of use are material under the provisions of the Building Regulations. Reference to Regulation 5 shows that control is only exercised over the creation of a dwelling, flat, hotel, boarding house, institutional building or public building from some other use or, where a building contains at least one dwelling, it has that number increased or decreased as a result of the change of use. Additionally, if the building originally fell within Classes I to VI of Schedule 2 (see section 2.3.3) and its new use meant that it could no longer be so classified, then that would also constitute a material change of use. Interestingly, not all the requirements of Schedule 1 apply to all types of change of use. Regulation 6 gives details of which requirements apply in each case and it is found, for example, that means of warning and escape (paragraph B1) applies in all cases whereas Part E (Resistance to the passage of sound) applies only in the case of the creation of a dwelling or flat.

2.5 Building Regulations – control by the local authority

Section 91 of the Building Act 1984 makes it a function of local authorities to enforce Building Regulations in their areas, and in England and Wales this function is exercised by District Councils. It is possible to submit plans and particulars, and have the work inspected on site by a private approved inspector in certain cases; however, the work of enforcement is still the responsibility of the local authority.

If it is proposed to use the local authority system of control then, in general, two options are available:

- the building notice procedure; *or*
- the full plans procedure.

2.5.1 Building notice procedure

A person intending to carry out building work to which the Regulations apply may give a building notice to the local authority *unless* the building is to be put to a 'relevant use'. This term is used to mean a workplace subject to Part II of the Fire Precautions (Workplace) Regulations 1997 or a use designated under the Fire Precautions Act 1971. This means that building work to most places of work must be dealt with under the full plans procedure, thus limiting the building notice procedure almost entirely to dwellings (for full details of designated buildings see Chapter 3).

There is no official form of building notice; however, it must be signed by, or on behalf of, the person intending to carry out the work and must contain a certain minimum amount of information. A typical notice is illustrated in Figure 2.1. It should be accompanied by a plan to a scale of not less than 1:1250. Full details of the building notice procedure are given in Regulations 12 and 13.

On receipt of the notice the local authority will usually acknowledge it and check that the necessary information has been provided. On commencement of the project the authority will carry out the usual inspections of the work in progress (see later in this chapter) and is entitled to request further plans or details to enable the building control functions to be discharged.

A building notice remains active for three years from the date of its receipt by the local authority. If the work does not commence within that period then the notice automatically lapses.

Building Regulations Building Notice

The Building Act 1984
The Building Regulations 2000

Building Control Services

Newtown Borough Council
Newtown, Leics LE99 9ZZ

This form is to be filled in by the person who intends to carry out building work or his/her agent. If difficulty is experienced in filling in this form please refer to the notes overleaf or consult the office indicated above.

1 **Applicant's details**
Name:
Address:

Postcode: Telephone: Fax:

2 **Agent's details** (if applicable)
Name:
Address:

Postcode: Telephone: Fax:

3 **Location of building to which work relates**
Address:
Postcode:

4 **Proposed work** (e.g. domestic extension, internal alterations, re-roofing etc.)
Description:

5 **Use of building** Domestic ☐ Non-domestic ☐ (tick box)

1. If new building or extension please state proposed use ------------------------------
2. If existing building please state present use --

A Building Notice may not be submitted for places of work or those that require a Fire Certificate

6 **Charges**

1. If Schedule 1 work please state the total number of dwellings and types - Total No. No. of types
2. If schedule 2 work please state floor area: m^2
3. If schedule 3 work please state the estimated cost of work excluding VAT: £

Building Notice Charge: £ plus VAT: £ Total: £

7 **Additional information:**
1. Number of storeys:
2. Number of bedrooms:
3. Means of water supply:

8 **Statement**
This notice is given in relation to the building work as described, is submitted in accordance with Regulation 13 and is accompanied by the appropriate payment.

This Building Notice shall cease to have effect three years from the date submitted to the Local Authority unless the work has previously commenced and the appropriate notification given to the Council.

Name: Signature: Date:

Fig. 2.1 Typical building notice.

2.5.2 *Full plans procedure*

This procedure may be used for all building types and the rules governing it are to be found in section 16 of the Building Act 1984 as supplemented by Regulation 14. On receipt of the details described in Regulation 14 the local authority must either pass or reject the proposals within five weeks, although this may be extended to two calendar months with the written permission of the applicant. Also with the applicant's written permission, the local authority may pass the plans conditionally or in stages and may impose conditions as to the deposit of future plans.

Normally, two copies of all the details must be submitted to the local authority; however, where the proposals involve Part B (Fire safety) provisions, then two further copies of the details must be submitted. This is in order that sufficient details are available to allow consultation with the fire authority. Additionally, if the proposals relate to a relevant use then the plans must be accompanied by a statement to that effect.

Work may be commenced as soon as plans are deposited provided that the local authority has been given at least two days' notice of commencement; however, any work carried out before the plans are passed will be at the applicant's own risk if the subsequent inspection of the plans shows them to be defective in some way.

2.5.3 *Building Regulations – supervision of works in progress by the local authority*

Once work has commenced, the local authority's powers of inspection and enforcement are activated and these are the same irrespective of the procedure which has been adopted for the submission.

2.5.4 *Building Regulations – notice requirements*

It is the duty of the person who is carrying out the work to give to the local authority certain notices to enable the specified inspections to be carried out. These notices should be in writing but it is possible with the agreement of the authority to use other means, such as the telephone.

Regulation 15 requires that the following minimum periods of notice be given to the local authority commencing on the day after the notice was served:

- before commencement – two days;
- before covering up any foundation excavation, foundation, damp-proof course, concrete or other material laid over a site, or covering up of any drain or private sewer to which the Regulations apply – one day.

Additionally, notice must be given to the local authority not more than five days after:

- the laying, haunching or covering up of any drain or sewer governed by Part H of Schedule 1;
- the completion of the building or works.

Where a building or part of a building is occupied before completion then notice must be given at least five days before occupation as well as within five days of completion.

Failure to submit the necessary notices leaves the person liable to enforcement action by the local authority, which could include him having to lay open, cut into or even pull down the work so that compliance with the Regulations could be demonstrated.

2.5.5 Building Regulations – contravening works

If building work is carried out which is contrary to the Regulations, the local authority has powers under section 36 of the Building Act to require its alteration or removal by serving a notice on the building owner. Failure to comply with the notice requirements within 28 days (or such longer period as allowed on application to a magistrates' court), may result in the work being carried out by the authority and the costs being recovered from the defaulter. There is a right of appeal to a magistrates' court, where the burden of proving non-compliance lies with the local authority. (See also section 7 of the Building Act – section 2.2.3).

A section 36 notice cannot be given where the works have been completed for more than 12 months or where the local authority has passed the plans (even if defective) and the work has been carried out in accordance with the plans.

As an alternative to the appeal procedure outlined above, the owner may adopt the procedure described in section 37 and obtain a written report from 'a suitably qualified person'. This is submitted to the local authority which *may* withdraw the section 36 notice and *may* pay the expenses incurred in obtaining the report. Also, the period of 28 days is automatically extended to 70 days so that there will be sufficient time available to obtain the report.

Of course, the local authority may reject the report and in this case it can be used as evidence in any appeal proceedings under section 40 of the Act.

2.5.6 Completion certificate procedure

If a local authority is satisfied, after taking all reasonable steps, that the relevant requirements of Schedule 1 to the Building Regulations have been met, then they are required to give a certificate to that effect (referred to as a completion certificate).

The issue of a completion certificate is mandatory in the following circumstances:

- Where the work involves buildings which are put to a relevant use. However, the certificate only needs to relate to the requirements of Part B (Fire safety) unless the next condition below is also met.
- Where the applicant has requested that a certificate be issued at the submission stage provided that the full plans procedure has been followed. In this case the certificate must relate to all the applicable requirements of the Building Regulations.

It should be noted that the certificate is evidence (but not *conclusive* evidence) that the requirements specified in the certificate have been complied with and that the local authority cannot be held liable for a contravention if they fail to give a certificate.

2.5.7 Building Regulations – relaxation of requirements

Section 8 of the Building Act, together with Regulation 11, gives local authorities the power to relax or dispense with any requirements of the Regulations which would be unreasonable in relation to a particular case.

Since the change to functional Regulations in 1985 most of the provisions are phrased in terms of 'reasonableness' or 'adequacy' and, therefore, the granting of a relaxation would mean acceptance of something which was unreasonable or inadequate. Consequently, this procedure is rarely used today and most variations from the Approved Document recommendations are dealt with by negotiations between the designer and the local authority or by applying for a determination.

If a dispute arises between the applicant and the local authority regarding the application of the Regulations to the proposals, then under the provisions of section 16(10)(a) of the Building Act, a joint application may be made to the Secretary of State for the Environment for a determination, and his decision is final.

2.6 Building Regulations – supervision otherwise than by local authorities

The powers exist within Part II (sections 47 to 58) of the Building Act 1984 for building work to be supervised by the private sector rather than by relying on the local authority control system described above. Detailed rules and procedures for this form of control may be found in the Building (Approved Inspectors etc.) Regulations 2000[14], where prescribed forms and notices are provided.

It has taken some considerable time for the approved inspector system to become fully operational even though the first approved inspector, the National House Building Council (NHBC), was approved on 11 November 1985. Their original approval related only to dwellings of not more than four storeys but this was later extended to include residential buildings up to eight storeys and this was further extended to include any buildings in 1998.

The approval of further corporate bodies as approved inspectors was held up by a number of factors, but was due mainly to the difficulty posed in obtaining the level of insurance cover which was required by the Department of the Environment. After a period of consultation new proposals for insurance requirements were agreed and these were implemented on 8 July 1996. At the same time the Construction Industry Council (CIC) was designated as the body for approving non-corporate inspectors, although initially the Secretary of State reserved the right to approve corporate bodies. From 1 March 1999 the CIC became responsible also for the approval of corporate approved inspectors.

In addition to the NHBC, three further corporate bodies were approved by the Secretary of State from 13 January 1997 and there has been a steady stream of others approved since that date, but at present, the NHBC remains the only body which deals with domestic construction (i.e. self-contained houses, flats and maisonettes).

2.6.1 Supervision by an approved inspector

The initial notice

If it is decided to engage an approved inspector, the developer and the inspector must jointly give to the local authority an initial notice in a prescribed form (see form 1 in Schedule 2 to the Approved Inspectors Regulations) which must be accompanied by a declaration signed by the insurer, that an approved scheme of insurance applies to the work. The notice must contain:

- a description of the work; *and*
- in the case of a new building or extension:

- a site plan showing the boundaries and location of the site;
- information regarding the proposals for the discharge of drains and private sewers, including the need to consult the sewerage undertaker; *and*
- a statement of any local legislation (see section 2.8 below) relevant to the work and the steps being taken to comply with it.

It is important that all the required information is provided in the prescribed form, because if the local authority is not satisfied that the notice contains sufficient information, they must reject it. The local authority has five working days in which to consider the notice and may only reject it on prescribed grounds (see Schedule 3 to the Building (Approved Inspectors etc.) Regulations 2000). If the local authority does not reject the initial notice within five working days (beginning on the day the notice is given) the authority is presumed to have accepted it without imposing requirements.

Once the notice has been accepted, or is deemed to have been accepted by the passing of five days, the approved inspector is responsible for supervising the work as regards:

- compliance with Regulations 4, 6 and 7 of the Building Regulations;
- compliance, where relevant, with Regulation 12 of the Approved Inspectors Regulations concerning the calculation of energy ratings for dwellings.

Independence of approved inspectors

The approved inspector must have no professional or financial interest in the work he supervises. Essentially, this means that he must be independent of the designer, builder or owner, unless the work is minor work This is defined as:

(1) the material alteration or extension of a one or two-storey house, provided that the house has no more than three storeys on completion of the work;
(2) the provision, extension or material alteration of a controlled service or fitting in any building;
(3) work involving the underpinning of any building.

Consulting the fire authority

Where it is intended to erect a building which is to be put to a relevant use and Regulation B1 (Means of escape in case of fire) also applies, the approved inspector is required, as soon as practicable, to consult the fire authority. He must give them sufficient plans to show that the work described in the initial notice will comply with Regulation B1. Additionally, he must allow the fire authority 15 working days to comment, and have regard to the views they express, before giving a plans certificate (see below) or final certificate (see below) to the local authority.

Plans certificate

The person carrying out the building work (i.e. the client of the approved inspector) may wish to have detailed plans of the work (or a part of it) certified as complying with the Regulations. In this case the approved inspector should be asked to supply a plans certificate. If he is satisfied with the plans he must give a plans certificate to the client and the local authority. This can be done at the time the initial notice is given, or later. Possession of a plans certificate can give the client valuable protection in the event that the initial notice is cancelled or ceases to be in force and no new initial notice is given or accepted, since the local authority cannot take enforcement action in respect of any work described in the plans certificate if it has been done in accordance with those plans.

Final certificate

When the work is complete the approved inspector must give the local authority a final certificate. The local authority may reject this only on the prescribed grounds, and they must do this within 10 working days. If a final certificate is rejected the initial notice ceases to be in force on the expiry of four weeks beginning with the day on which the rejection is given.

As in the case of a plans certificate, a final certificate need not relate to all the work specified in an initial notice.

Cancellation or lapse of initial notice

Where an initial notice is given for the erection, extension or material alteration of a building, and the building or extension or any part which has been materially altered is subsequently occupied and no final certificate is given, the initial notice will cease to have effect.

For most buildings there is a period of grace of eight weeks from the date of occupation before the initial notice lapses, but in the case of a building to be put to a relevant use the period is four weeks. Similarly, if the initial notice relates to a material change of use and no final certificate is given and the change of use takes place, the initial notice will cease to have effect eight weeks after the change of use takes place.

Once the initial notice has ceased to have effect, the approved inspector will be unable to give a final certificate and the local authority's powers to enforce the Building Regulations can revive (see below). Local authorities can, however, extend the appropriate period of grace either before or after it expires, and may wish to if they are reasonably confident that a final certificate will be given soon.

If an approved inspector for any reason cannot continue to supervise work for which he has given an initial notice he must inform his client and the local authority by cancelling the initial notice. If the client becomes aware that the approved inspector is unable to continue supervising the work he must cancel the initial notice himself. Alternatively, it is possible for the person carrying out the work to give a new initial notice jointly with a new approved inspector, provided that the new notice is accompanied by an undertaking by the original approved inspector that he will cancel the earlier notice as soon as the new notice is accepted.

When an initial notice ceases to be in force and another approved inspector is not engaged, the local authority becomes responsible for enforcing the Regulations in relation to any work which has been carried out and for which a final certificate has not been given. In that event the person carrying out the building work must provide the local authority on request with plans of the building work so far carried out. Additionally, the local authority may require that the work be cut into, layed open or pulled down so that they may ascertain whether any work not covered by a final certificate contravenes the Regulations. If it is intended to continue with partially completed work, the local authority must be given sufficient plans to show that the work can be completed without contravention of the Building Regulations. A fee, appropriate to that work, will also become payable.

Change of person intending to carry out work

Under section 51C of the Building Act, an approved inspector and a person who proposes to carry out work in succession to the person who gave the initial notice may jointly give a written notice to that effect to the local authority. The initial notice is then treated as having been given by the approved inspector and the new person intending to carry out the work.

Contravention of Building Regulations

Unlike a local authority an approved inspector has no power to enforce the Regulations. He may, however, inform his client by written notice if he believes that any work being carried out under his supervision contravenes the Regulations. If there is a failure to remedy the alleged contravention within three months he is obliged to cancel the initial notice, and must give the client and the local authority a cancellation notice in the prescribed form. This notice must specify the nature of the contravention unless a further initial notice relating to the work has been given and accepted.

Dealing with variations to the work

Where it is proposed to vary work which is the subject of an initial notice (e.g. building six units instead of five on a site) the person carrying out the building work and the approved inspector should give an amendment notice to the local authority. There is a prescribed form for an amendment notice (form 2 in Schedule 2 to the Approved Inspectors Regulations). It must contain the information which is required for an initial notice (see section 2.6.1) plus either:

a) a statement to the effect that all plans submitted with the original notice remain unchanged; or
b) copies of all the amended plans with a statement that any plans not included remain unchanged.

The local authority has five working days in which to consider the notice and they may only reject it on prescribed grounds. The procedure is identical to that for acceptance or rejection of an initial notice (see section 2.6.1).

Duration of validity of initial notices and plans certificates

A local authority may cancel an initial notice if the work does not appear to have started within three years, beginning on the date that the notice was accepted or deemed to have been accepted. A local authority may likewise rescind its acceptance of a plans certificate after three years if the work does not appear to have begun.

2.7 The Building Act and means of escape – additional provisions

2.7.1 *Building Regulations and the 'linked powers'*

A number of statutory controls are exercised by local authorities and these are activated by a Building Regulation submission (the so-called 'linked powers') . The Building Act 1984 contains a number of these but most have been overtaken by changes to the Building Regulations which now incorporate these original Building Act provisions. From the viewpoint of means of escape the most significant provision is contained in section 24 which deals with the necessity to provide:

> 'such means of ingress and egress and passages or gangways as the authority, after consultation with the fire authority, deem satisfactory, regard being had to the purposes for which the building is intended to be, or is, used and the number of persons likely to resort to it at any one time.'

Clearly, the intention of section 24 is that the relevant buildings should be provided with adequate means of escape.

Not all buildings are covered by this requirement since it refers only to:

- a theatre, and a hall or other building used as a place of public resort;
- a restaurant, shop, store or warehouse which employs more than 20 people and to which members of the public are admitted;
- a club licensed to serve intoxicating liquor (i.e. registered under the Licensing Act 1964[15]);
- schools which are not exempt from the Building Regulations; *and*
- a church, chapel or other place of public worship, but this does not include a private house to which members of the public might be admitted only occasionally. Also excluded from control are any churches and chapels which were so used before 1 October 1937 (i.e. the date of commencement of the Public Health Act 1936[16]).

Section 24 has its origins in pre-war public health legislation and has, to all intents and purposes, been superseded by paragraph B1 (means of escape) of the Building Regulations, since it does not apply to any building or extension which is covered by B1. Furthermore, any local Acts of Parliament (see below) which impose similar requirements to section 24 are also superseded by B1.

2.7.2 *Means of escape in certain high buildings*

Section 72 enables a local authority, after consultation with the fire authority, to require adequate means of escape in an existing *or* proposed building which has more than two storeys and contains floors exceeding 20 ft (6.1 m) above the surface of the street or ground on any side of the building. Not all building types are covered by this section since it refers only to those which are:

- let into flats or tenement dwellings;
- used as inns, hotels, boarding-houses, hospitals, nursing homes, boarding-schools, children's homes or similar institutions;
- used as restaurants, shops, stores or warehouses, where there are upper floors

which contain sleeping accommodation for people employed on the premises.

The local authority is empowered to serve a notice on the owner or developer (in the case of a proposed building), to carry out those works which are necessary in order to provide an adequate means of escape.

Regarding proposed buildings, Section 72 does not apply if paragraph B1 of the Building Regulations imposes a requirement. Additionally, for existing buildings, the local authority cannot apply the provisions of section 72 to a building which has a valid fire certificate provided under the Fire Precautions Act 1971 or is a workplace subject to the provisions of the Fire Precautions (Workplace) Regulations 1997 (see Chapter 11).

2.8 Local Acts of Parliament

Although the Building Act 1984 attempted to rationalise the main controls over buildings, there are in fact numerous pieces of local legislation with the result that many local authorities have special powers relevant to building control. Where a local Act is in force, its provisions must also be complied with, since many of these pieces of legislation were enacted to meet local needs and perceived deficiencies in national legislation. The Building Regulations make it clear that local enactments must be taken into account.

With the growth and development of building regulation control over fire precautions in particular, it is likely that most of the current local legislation is now outdated or has been superseded by the Building Regulations. In fact, some local enactments already contain a statutory bar which gives precedence to Building Regulations. However it is a wise precaution always to seek the advice of the building control body as to the existence of any local legislation, since local authorities are obliged by section 90 of the Building Act to keep a copy of any local Act provisions and these must be available for public inspection free of charge at all reasonable times.

A full list of local Acts of Parliament (excluding the London Building Acts) which are relevant to fire safety in buildings is given in Table 2.1, where it will be seen that the most common local provisions relating to building control are as follows.

Special fire precautions for basement garages or for large garages

The usual provision is that if a basement garage for more than three vehicles or a garage for more than 20 vehicles is to be erected, the local authority can impose access, ventilation and safety requirements.

Table 2.1 Local Acts of Parliament. (Researched and compiled by Carillion Specialist Services Ltd, a corporate approved inspector under the Building Regulations, and reproduced with their kind permission.)

The sections marked with an asterisk (*) in the third column are applicable where there are matters to be satisfied by a developer as part of a submission under Building Regulations. County Acts usually, but not inevitably, apply across a whole county. References to district authorities in the list merely illustrate those which may have been consulted, and do not necessarily constitute a complete list of districts within that county. However, where it is known that a particular section applies only in a particular district this is indicated. The DTLR is developing proposals to repeal redundant provisions of local Acts but this will take some time to achieve.

Local Act	Relevant sections	Sections applicable
County of Avon Act 1982	s.7 Parking places, safety requirements	*
• Bristol City Council		
Berkshire Act 1986	s.28 Safety of stands	
• Newbury District Council	s.32 Access for fire brigade	*
	s.36 Parking places, safety requirements	*
	s.37 Fire precautions in large storage buildings	*
	s.38 Fire precautions in high buildings	*
• Reading Borough Council		
Bournemouth Borough Council Act 1985	s.15 Access for fire brigade	*
	s.16 Parking places, safety requirements	*
	s.17 Fire precautions in certain large buildings	*
	s.18 Fire precautions in high buildings	*
	s.19 Amending s.72 Building Act 1984	
Cheshire County Council Act 1980	s.48 Parking places, safety requirements	*
• Chester City Council	s.50 Access for fire brigade	*
• Warrington Borough Council	s.49 Fireman switches	
	s.54 Means of escape, safety requirements	
County of Cleveland Act 1987		
• Stockton-on-Tees Borough Council	s.5 Access for fire-fighting	*
	s.6 Parking places, safety requirements	*
	s.15 Safety of stands	
• Middlesbrough Borough Council		
Clwyd Act 1985	s.19 Parking places, safety requirements	*
• Borough of Rhuddlan	s.20 Access for fire brigade	*
• Colwyn District Council		
Cornwall Act 1984		
• North Cornwall District Council		
• Caradon District Council		
Croydon Corporation Act 1960	s.93/94 Buildings of excess cubic capacity	*
	s.95 Buildings used for trade and for dwellings	
Cumbria Act 1982		
• Barrow Borough Council	s.23 Parking places, safety requirements	*
• Carlisle City Council	s.25 Access for fire brigade	*
	s.28 Means of escape from certain buildings	

Cont.

Table 2.1 Continued.

Local Act	Relevant sections	Sections applicable
Derbyshire Act 1981	s.16 Safety of stands	*
• Borough of High Peak	s.23 Access for fire brigade	*
	s.24 Means of escape from certain buildings	*
	s.28 Parking places; safety requirements	*
	s.25 Fireman switches	
Dyfed Act 1987		
• South Pembrokeshire District Council	s.46 Safety of stands	
	s.47 Parking places; safety requirements	*
	s.51 Access for fire brigade	*
• Carmarthen District Council		
East Ham Corporation Act 1957		
• Newham London Borough Council	s.54 Separate access to tenements	*
	s.61 Access for fire brigade	*
East Sussex Act 1981	s.34 Fireman switches	
• Hastings Borough Council	s.35 Access for fire brigade	*
Essex Acts 1952 and 1958 (GLC areas formerly in Essex)		
Essex Act 1987		
• Uttlesford District Council	s.13 Access for fire brigade	*
Exeter Act 1987		
Greater Manchester Act 1981		
• Trafford Metropolitan Borough Council	s.58 Safety of stands	
	s.61 Parking places, safety requirements	*
	s.62 Fireman switches	
	s.63 Access for fire brigade	*
	s.64 Fire precautions in high buildings	*
	s.65 Fire precautions in large storage buildings	*
	s.66 Fire and safety precautions in public and other buildings	*
• Manchester City Council		
Hampshire Act 1983		
• Southampton City Council	s.11 Parking places, safety requirements	*
	s.12 Access for fire brigade	*
	s.13 Fire precautions in certain large buildings	*
Hastings Act 1988		
Hereford City Council Act 1985	s.17 Parking places; safety requirements	*
	s.18 Access for fire brigade	*
Humberside Act 1982		
• Gt Grimsby Borough Council	s.12 Parking places, safety requirements	*
	s.13 Fireman switches	
	s.14 Access for the fire brigade	*
	s.15 Means of escape in certain buildings	

Cont.

Table 2.1 Continued.

Local Act	Relevant sections	Sections applicable
Isle of Wight Act 1980	s.32 Access for fire brigade	*
• (Part VI)	s.31 Fireman switches	
	s.30 Parking places; safety requirements	*
Kent Act 1958 (GLC areas formerly in Kent)		
County of Kent Act 1981		
• Canterbury City Council	s.51 Parking places, safety requirements	*
• Rochester City Council	s.52 Fireman switches	
	s.53 Access for fire brigade	*
Lancashire Act 1984	s.31 Access for fire brigade	*
Leicestershire Act 1985	s.21 Safety of stands	
• Leicester City Council	s.49 Parking places, safety requirements	*
	s.50 Access for fire brigade	*
	s.52 Fire precautions in high buildings	*
	s.53 Fire precautions in large storage buildings	*
• North West Leicestershire District Council	s.54 Means of escape, safety requirements	
County of Merseyside Act 1980		
• Liverpool City Council	s.20 Safety of stands	
	s.48/49 Means of escape from fire	
	s.50 Parking places, safety requirements	*
	s.51 Fire and safety precautions in public and other buildings	*
	s.52 Fire precautions in high buildings	*
	s.53 Fire precautions in large storage buildings	*
	s.54 Fireman switches	
	s.55 Access for fire brigade	*
• Borough of Wirral		
Middlesex Act 1956 (GLC areas formerly in Middlesex)	s.33 Access for fire brigade	*
Mid Glamorgan County Council Act 1987		
• Merthyr Tydfil	s.9 Access for fire brigade	*
• Taff-Ely Borough Council		
Nottinghamshire Act 1985		
• City of Nottingham		
Plymouth Act 1987		
Poole Act 1986	s.10 Parking places, safety requirements	*
	s.11 Access for fire brigade	*
	s.14 Fire precautions in certain large buildings	*
	s.15 Fire precautions in high buildings	

Cont.

Table 2.1 Continued.

Local Act	Relevant sections	Sections applicable
County of South Glamorgan Act 1976 • Cardiff City Metropolitan District Council	s.27 Safety of stands s.48/50 Underground parking places s.51 Means of escape for certain buildings s.52 Fireman switches s.53 Precautions against fire in high buildings	 * *
South Yorkshire Act 1980 • City of Sheffield Metropolitan District Council	s.53 Parking places, safety precautions	*
• Barnsley Metropolitan Borough Council	s.54 Fireman switches	
• Rotherham Metropolitan Borough Council	s.55 Access for fire brigade	*
• Doncaster Metropolitan Borough Council		
Staffordshire Act 1983 • Staffordshire Moorlands District Council	s.25 Parking places, safety precautions s.26 Access for fire brigade	* *
Surrey Act 1985 • Guildford Borough Council • Spelthorne Borough Council	s.18 Parking places, safety requirements s.19 Fire precautions in large storage buildings s.20 Access for fire brigade	* * *
Tyne & Wear Act 1980	s.24 Access for fire brigade	*
West Glamorgan Act 1987 • City of Swansea • Neath Borough Council	s.43 Parking places, safety precautions	*
West Midlands Act 1980 • Birmingham City Council	s.39 Safety of stands s.44 Parking places, safety requirements s.45 Fireman switches s.46 Access for fire brigade s.49 Means of escape from certain buildings	 * *
West Yorkshire Act 1980 • Bradford City Council	s.51 Fireman switches	
Worcester City Council Act 1985		

Fire precautions in high buildings or for large storage buildings

There must be adequate access for the fire brigade in certain high buildings. A high building is often defined as one in excess of 18.3 m (60 ft) and the local authority must be satisfied with the fire precautions. Additionally, it may impose conditions, e.g. fire alarm systems, fire brigade access, etc. (In many cases these requirements have been superseded by Part B of Schedule 1 to the Building Regulations 2000.) Large storage buildings in excess of 14,000 m^3 can be required to be fitted with sprinkler systems by some local Acts.

Extension of means of escape provisions

Section 72 of the Building Act 1984 (see section 2.7.2) is a provision under which the local authority can insist on the provision of means of escape where there is a storey which is more than 20 ft above ground level in certain types of buildings, e.g. hotels, boarding houses, hospitals, etc. Local enactments replace the 20 ft by 4.5 m and make certain other amendments to the national provisions.

Safety of stands at sports grounds

In many areas, local Acts impose controls over the safety of stands at sports grounds. Again much of this local legislation has been largely superseded by the provisions of the Fire Safety and Safety of Places of Sport Act 1987.

2.9 The London Building Acts

Although much of the content of the London Building Acts was repealed when the Building (Inner London) Regulations 1985[17] came into force, a number of provisions were retained which apply only to building work in Inner London (and are somewhat analagous to the Local Acts of Parliament encountered in other parts of England and Wales – see section 2.8 above). These powers are usually administered by the local authority building control department of the relevant Inner London Borough. Those Boroughs at present include:

- Islington;
- Hackney;
- Tower Hamlets;
- Southwark;
- Lambeth;
- Wandsworth;
- Hammersmith;
- Kensington and Chelsea;
- Camden;
- The Corporation of the City of London.

The most important provisions relating to fire safety are contained in the

London Building Acts (Amendment) Act 1939[18]. These provisions have been further amended by both the 1985 Regulations and the Building (Inner London) Regulations 1987[19], and are as follows.

Section 20 – Buildings in excess height and cubical content

This section applies to buildings which are to be erected and:

(1) have a storey at a height greater than 30 m, or 25 m if the area of the building exceeds 930 m^2; *or*
(2) are warehouses or are used for trade or manufacture and have a volume exceeding 7100 m^3 (unless fire-resisting division walls are provided as defined in subsection (2) of section 20 to limit the volume to 7100 m^3).

On receipt of a notice or plans from the owner or occupier, the local authority, after consulting the London Fire and Civil Defence Authority, may impose conditions for the provision and maintenance of:

(1) fire alarms;
(2) automatic fire detection systems;
(3) fire extinguishing appliances and installations;
(4) effective means of removing smoke in case of fire; *and*
(5) adequate means of access to the interior, exterior and site of the building for fire brigade personnel and appliances.

Items (1) and (3) above do not apply to any building for which a fire certificate issued by the Health and Safety Executive is required. This includes factories where highly flammable materials are stored or used and premises used for certain hazardous processes and products (see the Fire Certificates (Special Premises) Regulations 1976[20]).

The local authority may also impose additional conditions in respect of any special fire risk area (as defined in section 20 subsection (2D)) for:

- restricting the use of any such area in the building; and
- the provision and maintenance of proper arrangements for lessening danger from fire in the building (so far as this is reasonably practicable).

Section 21 – Uniting of buildings

The local authority's consent must be sought for the uniting of buildings:

(1) by making an opening in a party wall or external wall separating buildings; *or*

(2) when buildings are connected by access without passing into the external air,

unless, when united, they are wholly within one occupation and as such comply with the London Building Acts.

If a building which has been previously united becomes separately occupied, notice must be given to the local authority, and the openings between the buildings must be stopped up unless the local authority agrees to their retention.

Section 34 – Means of escape in case of fire in new buildings

On receipt of a building notice or plans which show the proposed means of escape in a new building, the local authority may refuse to approve the means of escape or may approve it subject to conditions. They have up to two months (or such longer period as may be agreed in writing) to come to a decision. The provisions apply to:

(1) certain public buildings;
(2) single-storey buildings exceeding 600 ft^2 (about 56 m^2) in area;
(3) buildings of more than one storey (not including storage basements) exceeding 1000 ft^2 (about 93 m^2) in area;
(4) buildings (other than single family dwellings) with a storey higher than 20 ft (6.1 m); *and*
(5) buildings in which more than 10 people are employed above the ground storey.

Curiously, the provisions apply to buildings which are not covered by Part B1 of the Building Regulations 1985. B1 only applied to certain houses, flats, offices and shops and was repealed when the 1991 Building Regulations came into force, Therefore, it would appear that in Inner London both Part B1 and section 34 apply to the buildings mentioned in (1) to (5) above.

2.9.1 Exemptions

It should be noted that a large number of buildings are exempt from the provisions of the sections listed above. Therefore, reference should be made to sections 149 and 150 of the London Building Acts (Amendment) Act 1939 where a list of exemptions may be found. These range from individual named buildings, such as the Mansion House, Guildhall and the offices and buildings of the Bank of England, to classes of buildings, such as railway stations and certain statutory undertakers' buildings.

2.10 Houses in multiple occupation

Under the Housing Act 1985 local authorities (LAs) have the power to require means of escape in case of fire in houses which are occupied by persons not forming a single household (known as houses in multiple occupation or HMOs). In the case of HMOs with at least three storeys and a floor area (all storeys combined) of at least 500 m^2, LAs have a duty to ensure that means of escape are adequate. Before exercising their power or performing their duty in respect of means of escape from fire in an HMO, LAs must consult the fire authority.

For new HMOs with six or fewer residents the guidance in Chapter 6, sections 6.2 to 6.3.8 (which deals with means of escape from single family dwellinghouses) is adequate. For details of the local authority's powers in relation to existing HMOs see Chapter 3. A summary of the technical requirements in such buildings is given in Chapter 7.

2.11 References

1 Building Act 1984, c.55
2 The Building Regulations 2000 (SI 2000/2531).
3 Prisons Act 1952, c.52.
4 *LPC Design guide for the fire protection of buildings*. Loss Prevention Council, 1996.
5 *Rickards* v. *Kerrier District Council* (1987) Construction Industry Law Letter 345; Construction Law Digest 4-CLD-04-26.
6 *Gorton Local Board* v. *Prison Commissioners* (1887).
7 Carillion Specialist Services Ltd, Westlink House, 981 Great West Road, Brentford, Middlesex.
8 The Education (Schools and Further and Higher Education) (Amendment) (England) Regulations 2001 (SI 2001/692).
9 Postal Services Act 2000 (Consequential Modifications to Local Enactments No.1) Order 2001 (SI 2001/648).
10 Explosives Act 1875, c.17.
11 Explosives Act 1923, c.17.
12 Nuclear Installations Act 1965, c.57; as amended by SI 1974/20056.
13 Ancient Monuments and Archaeological Areas Act 1979, c.46.
14 The Building (Approved Inspectors etc.) Regulations 2000, (SI 2000/2532).
15 Licensing Act 1964, c.26.
16 Public Health Act 1936, c.49.
17 The Building (Inner London) Regulations 1985 (SI 1985/1936).
18 The London Building Acts (Amendment) Act 1939, Chapter xcvii.
19 The Building (Inner London) Regulations 1987 (SI 1987/798).
20 The Fire Certificates (Special Premises) Regulations 1976 (SI 1976/2003).

Chapter 3

Buildings in Use – the Statutory Requirements

3.1 Introduction

In the previous chapter it was shown that the principal legal controls over means of escape in new and altered buildings are through the provisions of the Building Act 1984 and the Building Regulations 2000. Therefore, if a building is constructed in accordance with the requirements of the Regulations there should be no need to comply with other legislation unless the building is further changed so that its means of escape becomes unacceptable. The main problem with this argument is that the Building Regulations do not allow for continuing control and therefore there would be no way of ensuring that the means of escape remained viable while the building was being used.

A study of the historical development of means of escape legislation shows that this was first applied to buildings in use due to the hazardous operations or unsafe conditions which often occurred in them. As we have shown in Chapter 1, the application of Building Regulations at the design and construction stage is a relatively new innovation and this has developed alongside the existing use-orientated legislation.

As an example of this dual approach, most places of work need to comply with both the Building Regulations (to ensure that they possess a safe means of escape initially) and the Fire Precautions Act and/or the Fire Precautions (Workplace) Regulations (so that the means of escape continues to be safely and effectively used during the life of the building). This situation involves complicated joint consultation between two different enforcing bodies in order that all legislation is satisfied.

The remainder of this chapter attempts to clarify the current legislative situation regarding buildings in use by describing:

- what requirements need to be met;
- when the requirements are applied to the building;

- who is the controlling authority and what are their powers of enforcement; *and*
- which design standards are applied and what is their legal status.

3.2 Buildings in use – certification, licensing and registration of premises

With regard to fire safety, there are a great many statutory requirements relating to different building uses, which are applied by different enforcement bodies at different times during the life of the building, and to different technical standards. Until 1997, when the Fire Precautions (Workplace) Regulations came into force, the most widely applied form of control over means of escape in buildings in use was through the provisions of the Fire Precautions Act 1971. This requires third party certification (by the local fire authority) for relevant buildings and is the most common form of building certification. Other forms of control have been developed piecemeal over the years to deal with perceived weaknesses in fire safety legislation and these apply certification, licensing or registration provisions to existing buildings with specific uses. The common factor connecting all such existing legislation is the need to apply to a third party for permission to carry on a particular building use. This will also involve some form of continuing control where the third party will inspect the premises at intervals to ensure compliance is maintained.

In recent years the emphasis has moved away from third party control into the realms of individual responsibility, and the most recent piece of fire safety legislation at the time of writing (the Fire Precautions (Workplace) Regulations 1997 (as amended)) imposes duties on employers and on others in control of workplaces, with regard to the provision of minimum fire safety standards in places of work. The future is likely to see a new Fire Safety Act repealing the Fire Precautions Act and imposing a general duty of fire safety care on employers, occupiers and owners of premises.

3.3 Fire certification – the Fire Precautions Act 1971 (as amended)

3.3.1 Control over means of escape by fire certification

All buildings to which the Act applies are required to have a fire certificate unless they are exempt. Fire certificates are issued by the local fire authority (county council or metropolitan fire authority), but the task of carrying out inspections and assessments of premises is delegated to local fire brigades. Section 6 of the Act *requires* that a fire certificate must specify:

- the use or uses of the premises;
- details of the means of escape which is provided;
- how the means of escape can be safely and effectively used at all material times;
- details of the means for fighting fire; *and*
- the method of giving warning in case of fire.

Additionally, the fire authority has *discretion* to impose further requirements such as:

- details of how the means of escape, fire-fighting equipment and warning systems are to be maintained;
- the method for ensuring that the means of escape will remain unobstructed during use;
- details of staff training programmes (with accompanying training records);
- limitations on the numbers of people present in the premises at any particular time; and
- any other precautions which may need to be observed in relation to the risk, in case of fire, to persons in the premises.

3.3.2 Fire certification and designated buildings

Fire certificates are only required for premises which are put to certain 'designated' uses and the Secretary of State is enabled to 'designate' uses from the following list contained in Section 1 of the Act:

- any purpose involving sleeping accommodation;
- institutions providing treatment or care;
- premises used for entertainment, recreation or instruction (including use as a club, society or association);
- teaching, training or research premises;
- premises accessible to the public, whether on payment or otherwise; *and*
- places of work.

Use as a single private dwellinghouse is specifically exempted by section 2 of the Act.

The original intention of the Act was that all the above uses (except private dwellinghouses) would be designated over a period of time, but to date only two designating orders have been made covering:

(1) hotels or boarding houses where sleeping accommodation is provided for more than six staff or guests (or some sleeping accommodation is provided above the first floor or below the ground floor); *and*

(2) factories, offices, shops and railway premises where more than 20 people are at work at any particular time, or more than 10 people work other than on the ground floor.

Therefore, it is clear that although some premises may be put to a designated use, they may also be exempt from the need to obtain a fire certificate by virtue of the numerical cut-offs mentioned above. Additionally, many types of premises are exempt from the need to obtain a fire certificate simply because they are not covered by the designating orders already made.

3.3.3 Fire certification – links with Building Regulations

Therefore, premises which are put to a designated use (and fall within the occupancy limitations outlined above) are required to have a fire certificate in accordance with section 1 of the Act, and occupiers who use premises without a certificate being in force (or without having applied for one) may leave themselves open to prosecution. It should be noted that the obligation to comply with the fire certification procedure usually falls on the occupiers of premises; however, in the case of multiple occupation this obligation now falls on the owners.

Clearly, there are two possible situations where the fire certification process becomes operative:

(1) where the building is to be put to a designated use for the first time and, therefore, does not have a fire certificate; *and*
(2) where the building already has a fire certificate (even though this may be under former legislation) and there is a material change in its condition necessitating the issue of a new or amended certificate.

In the first case, the need for a certificate may arise because the building has just been constructed. Equally, if a building has its use changed to one of those which are designated under the Act then it will also need a fire certificate. Reference to Chapter 2 will show that most new buildings are also required to comply with the Building Regulations 2000. Additionally, existing buildings come under Regulation control when they are subject to a material change of use (see section 2.4.4).

In the second case, the need also to comply with Building Regulation requirements will depend on the degree of change which is taking place and whether this is deemed to be a 'material alteration' under the provisions of Building Regulation 3 (see section 2.4.3).

It is important to realise that where it is proposed to carry out building works which are subject to Regulation control then the building control body (local authority or Approved Inspector) should be approached first and a Building Regulation submission made (if using a local authority) or initial notice submitted (if using an approved inspector). (In fact, if an initial approach is made to the fire authority, the applicant will be directed to a building control body.) Under section 16 of the Fire Precautions Act the local authority has a duty to consult the fire authority, before passing or rejecting a submission, if it appears that the new (or changed) use of the building would be a designated use. This allows the fire authority to comment on fire safety matters covered by Building Regulations and permits early agreement with the applicant on fire safety matters not addressed by Building Regulations, such as fire-fighting equipment.

Similar consultation arrangements exist where an approved inspector is engaged to carry out the building control function. Where it is intended to erect a building which is to be put to a relevant use (i.e. a workplace subject to Part II of the Fire Precautions (Workplace) Regulations 1997 or a use designated under the Fire Precautions Act 1971) and Regulation B1 (Means of escape in case of fire) also applies, the approved inspector is required, as soon as practicable, to consult the fire authority. He must give them sufficient plans to show that the work described in the initial notice will comply with Regulation B1.

Additionally, he must allow the fire authority 15 working days to comment and must have regard to the views they express, before giving a plans certificate (see section 2.6.1) or final certificate (see section 2.6.1) to the local authority.

In the past there has been some confusion regarding the legal status of any advice offered by the fire authority during the consultation process. Paragraph 6.8 of DTLR Circular 07/00 dated 13 October 2000, which accompanied the issue of the 2000 Approved Inspector Regulations, clarifies this issue:

> 'There is no longer an implication that the fire authority have an authoritative view on the compliance of building work with Part B, although it is of course open to them to offer informal views on that matter. The primary object of the consultation is to provide an opportunity for the approved inspector and the fire authority to reach mutually compatible views on whether plans of the building work are satisfactory from the standpoints of the building regulations and of fire precautions legislation.'

3.3.4 *Fire certification – use of the statutory bar*

Since it is undesirable to have to comply with more than one set of standards when carrying out building work, section 13 of the Fire Precautions Act imposes a 'statutory bar' on the fire authority. This prevents them from making the issue of a fire certificate conditional on alterations to the means of escape, *if when the building was erected the Regulations imposed means of escape requirements.* However, the statutory bar does not apply if the means of escape is inadequate by virtue of matters which did not have to be shown in connection with the deposit of plans under the Regulations. For example, the statutory bar might not apply in the case of a new factory which had its means of escape designed for normal use, if on occupation it was used for a process which involved a high risk.

Additionally, the statutory bar does not apply to any place of work covered by the Fire Precautions (Workplace) Regulations. Therefore these Regulations allow the fire authority to comment on the proposals for:

- equipping the workplace with fire-fighting equipment and fire detectors and alarms;
- ensuring that fire-fighting equipment is easily accessible, simple to use and indicated by appropriate signs;
- making sure that properly trained employees are nominated to implement the necessary fire safety measures;
- making sure that adequate arrangements are made regarding contact with the emergency services;
- providing suitable emergency routes and exits;
- arranging for a system of maintenance to ensure that all equipment is kept in good working order.

It is the duty of every fire authority to enforce the Fire Regulations in their area and they have powers to inspect premises at any time to ensure compliance with the Regulations. (For a detailed treatment of the practical considerations contained in the Fire Regulations see Chapter 11.)

3.3.5 Fire certification – application and enforcement

If it is decided that a fire certificate is needed then application is made to the fire authority in the prescribed form in accordance with the requirements of section 5 of the Act. On receipt of the application the fire authority may request that specified plans be supplied within a time limit. If plans are not forthcoming they may assume that the application has been withdrawn. Once this stage has been reached the fire authority is under a duty to carry out an inspection of the relevant premises and if on doing this they can satisfy themselves that:

- the physical means of escape which is provided;
- the means for securing the safe and effective use of the means of escape;

- the means for fighting fire (for use by the occupants), provided within the building; *and*
- the method of giving warning in case of fire;

are such as may reasonably be required in the circumstances then they must issue a fire certificate.

If they are not satisfied with the application after carrying out the inspection, they must serve a notice on the applicant informing him of the steps which need to be taken in order to achieve the necessary standard. This may, of course, involve the making of alterations to the premises and before requiring alterations to be made to the building the fire authority is obliged under section 17 of the Act to consult the relevant building control authority (since the alterations may also need to comply with the Building Regulations). The notice will specify the time (which may be extended – see section 9 on appeals) in which the alterations are to be carried out. Once the final time limit has been exceeded without the work having been completed, the certificate is deemed to have been refused and further use of the premises is a contravention of the Act, making the occupier guilty of an offence.

Once a fire certificate is in force the fire authority has the powers to inspect the premises at 'any reasonable time' to ensure that the occupier is adhering to the conditions which have been laid down. They may also inspect the fire certificate and obtain necessary facilities and assistance from responsible people at the premises; intentional obstruction of an inspector operating under the Act is an offence.

Sometimes the conditions of use of a building are changed so that the content of the certificate is rendered inadequate. For example, there may be a slight change to the internal layout which affects the means of escape or there may be a proposal to extend or structurally alter the premises. Section 8 requires that the occupier notify the fire authority if such alterations are contemplated, and if they are carried out without this notification he is guilty of an offence. It is then up to the fire authority to inform the occupier of the steps needed to prevent the inadequacy and once these have been taken the fire authority must either amend the certificate or issue a new one. Again, the fire authority must consult the building control authority before serving any notice requiring alterations to be made. The fire authority may charge a fee for issuing or amending a certificate but this must not exceed the cost of work reasonably done and must not include the costs of carrying out the inspections.

Where a fire authority is of the opinion that the fire risk within certain premises is so serious that their use should be prohibited or restricted until the risk has been reduced to a reasonable level, they are able to serve a prohibition notice on the occupier (Fire Precautions Act, section 10). The notice can have immediate effect if the authority considers that the risk of personal injury is imminent, otherwise the effect can be deferred pending the carrying out of remedial works. Section 10A of the Act provides for an appeal procedure to the courts. This does not permit the suspension of the prohibition notice until the appeal is disposed of, unless the court so directs.

3.3.6 Control of premises which are exempt from fire certification under the Fire Precautions Act

As mentioned above, some designated premises are exempt from the need for a fire certificate because they are below certain numerical thresholds related to their occupancy (see section 3.3.2).

Powers also exist which enable fire authorities to exempt certain low risk commercial and industrial premises from the requirement to have a certificate. Under section 5A of the Fire Precautions Act fire authorities are given discretionary powers to exempt relevant premises if they fall within prescribed criteria. These criteria are set out in the Fire Precautions (Factories, Offices, Shops and Railway Premises) Order 1989[1]. The guidance contained in the Order consists of factors to be borne in mind by the fire authorities and does not contain definitive rules which must be followed. Consequently, there has been reluctance on the part of fire authorities to exercise their discretionary powers of exemption, and differences of interpretation are an obvious possibility. Additionally, where a fire authority has granted exemption under section 5A they may, if they think fit, withdraw the exemption at any time (section 5B).

Therefore, although it is possible for designated premises to be exempt from the need for a certificate by virtue of the Secretary of State's powers under section 1(3) of the Act (i.e. they fall below the numerical thresholds) or because they fall within the scope of the fire authorities' discretion to exempt under section 5A, they may still be controlled under the powers contained in section 9A of the Act.

This section places a general statutory duty on the occupiers of exempt premises (no matter how they are exempted) to provide them with such means of escape and means for fighting fire as are reasonable in the circumstances. Because of the way in which the law has been implemented, at present the premises covered by section 9A are mostly small offices, shops and factories.

Occupiers of such premises who contravene this duty commit an offence; however practical guidance is available in the form of a code of practice[2] issued by the Secretary of State. Use of the code of practice is not obligatory and failure to comply with its recommendations does not automatically involve any civil or criminal liability; however the code can be relied upon by either party in any proceedings about an alleged contravention of the statutory duty. If an occupier proves that he has complied with the recommendations of the code, in any proceedings which are brought against him he can rely on this 'as tending to negative liability'. Conversely, failure to comply with the code may be relied on by the fire authority 'as tending to establish liability'.

Where a fire authority considers that the statutory duty to provide adequate means of escape has been contravened they may serve an improvement notice on the occupier specifying the steps which need to be taken to remedy the defects within a specified time. There is a right of appeal under section 9E, which has the effect of suspending the operation of the improvement notice.

It is likely that these provisions for controlling exempt premises will fall into disuse since the Fire Precautions (Workplace) Regulations 1997 now disapply section 9A for all buildings to which they apply and there are few, if any, workplaces that are not covered by the Fire Regulations.

3.4 Certification – other statutory controls

3.4.1 The Fire Certificates (Special Premises) Regulations 1976[3]

These Regulations define those premises for which a fire certificate will be required if certain quantities of hazardous substances are used or stored or certain hazardous activities are carried out. Schedule 1 to the Regulations gives details of the materials and processes (e.g. any premises at which oxygen is manufactured or stored in quantities exceeding 135 tonnes).

The Regulations apply to the whole site and not to individual buildings or plant (except for licensed explosive factories). An application for a fire certificate should be made to the Health and Safety Executive (HSE) (from whom advice can be sought) when the use or storage of a relevant substance reaches the specified limit.

A fire certificate may impose such conditions as the HSE consider appropriate and may include:

- the provision and maintenance of adequate means of escape;
- limitations on the numbers of persons on the premises at any one time;
- limitations on the quantities and disposition of hazardous materials on the site;
- requirements for staff instruction and training in what to do in case of fire including ensuring that training records are kept.

3.4.2 The Safety of Sports Grounds Act 1975[4]

This Act (as amended by the Fire Safety and Safety of Places of Sport Act

1987[5]) requires that designated sports grounds which accommodate more than 10,000 spectators (or 5000 spectators if subject to the Safety of Sports Grounds (Accommodation of Spectators) Order 1996[6]) be covered by a safety certificate issued by the local authority. The certificate will contain terms and conditions deemed appropriate by the authority such as:

- the maximum number of spectators permitted to be present;
- details of entrances and exits;
- details of crash barriers;
- means of escape in case of fire.

When applying for a safety certificate, the procedure to be adopted is governed by the Safety of Sports Grounds Regulations 1987[7]. Notices must be served on the local authority, the chief officer of police and the fire authority. Once a certificate has been issued by the local authority they are under obligation to inspect any designated ground at least once a year.

Where the local authority considers that there is a serious risk to spectators, it may serve a prohibition notice on the person in charge of the premises. This will have the immediate effect of prohibiting or restricting admission to the ground.

Guidance on the design and management of new grounds and on measures that may be taken to improve safety at existing grounds is available in the *Guide to Safety at Sports Grounds*[8]. This document is a distillation of many years of research and experience of the safety management and design of sports grounds. It has no statutory force but many of its recommendations will be given force of law at individual grounds by their inclusion in safety certificates, and most local authorities rely on the advice contained in the guide.

3.5 Licensing controls

3.5.1 Introduction

Traditionally, the control of premises used for the consumption of alcohol or for public entertainment was a reflection of society's views concerning the need to prevent the occurrence of disorderly or criminal conduct. More recently, greater emphasis has been placed on the physical dangers that may arise where large numbers of people are present (possibly adversely affected by the consumption of alcohol). This includes danger from fire and the possibility of poor construction or lack of maintenance in the premises.

As a result, the legislation has developed in a piecemeal fashion as a reflection of the changing needs of society, and now is scattered throughout a bewildering body of statutes. In most cases where the need for a licence exists, it is illegal to use the premises without such a licence.

3.5.2 The Licensing Act 1964[9]

This Act covers premises used for the sale of alcohol whether consumed on the premises (such as public houses, clubs, hotels and restaurants) or purchased for consumption in some other place (such as from an off-licence).

Section 3 of the Act provides that:

> 'licensing justices may grant a justices' licence to any such person, who is not disqualified under this or any other Act for holding a justices' licence, as they think fit or proper.'

Licences are granted by the licensing justices at their 'Licensing Sessions' (i.e. the General Annual Licensing Meeting and at least four other transfer sessions held at regular intervals during the year). The General Annual Licensing Meeting is more commonly known as the Brewster Sessions and is held in the first fortnight of February each year.

Anyone applying for:

- the grant of a new licence; *or*
- the transfer of a licence (i.e. from one person to another); *or*
- removal of a licence (i.e. the transfer of the licence from one set of premises to another);

must give written notice at least 21 days before the day of the licensing sessions to the clerk to the justices, the chief officer of police, the proper local authority and the relevant fire authority (in the case of a new licence or a removal). The notice must be accompanied by suitable plans of the premises if it concerns a new licence or a removal. The Act does not lay down any precise requirements as to the form which the plans should take. Therefore, enquiries should be made of the clerk to the justices to determine whether any local regulations have been made which specify items such as scale, colouring up, etc.

The justices may require the applicant for a licence to attend the licensing session in person and they may also wish to inspect the premises. During the hearing of the licensing application the applicant or his representative will be expected to give evidence in support of the application. Anyone objecting to the licence is also entitled to give evidence and both the applicant and any objector(s) may be cross-examined. In these circumstances, fire officers or representatives of the building control body may need to give evidence regarding the acceptability of the means of escape and other fire safety provisions to be provided in the premises.

The justices are entitled to attach such conditions as they see fit when granting a new on-licence (i.e. a licence for premises where alcohol is consumed on the premises). These conditions normally relate to the way the alcohol will be served (e.g. they may limit the sale of alcohol to a particular part of the premises) and not to the general state or nature of the premises. Where the subject matter is not thought to be appropriate for the imposition of a condition on the licence, the justices may require undertakings to be given. For example, they may request that building work required under Building Regulations or work required by the fire officer is carried out before the premises are used. Clearly, such work will need to comply with the relevant legislation, and the usual guidance documents (i.e. Approved Documents and British Standards) will apply.

3.5.3 The Local Government (Miscellaneous Provisions) Act 1982[10]

Under this Act certain types of public entertainment must be licensed by the local authority. The entertainment covered includes:

- public dancing or music or any other public entertainment of a like kind;
- any public contest, exhibition or display of boxing, wrestling, judo, karate or any similar sport.

Music which is provided in a place of public religious worship (or which is performed as an incident of a religious meeting or service) is excluded from the provisions for licensing as are any of the entertainments listed if held in a pleasure fair.

An application for a licence (with the appropriate fee) must be made to the relevant local authority, the chief officer of police and the fire authority at least 28 days before the entertainment is due to take place. The information that needs to be provided with the application will be specified by the local authority, since they will need to be satisfied that:

- the safety (including safety in the event of fire) of both performers and spectators is ensured;
- adequate access arrangements are made for vehicles belonging to the emergency services;
- adequate sanitary arrangements are being made; *and*
- disturbance to the neighbourhood will be prevented.

Whilst any conditions imposed on the licence must be only for the general purposes listed above, the very generality of these purposes provides considerable scope for local authorities to insist on lengthy and complex requirements as a condition of granting a licence.

Full licences are granted for periods of up to one year (annual licences). They may also be granted for one or more specific occasions (occasional licences) and it is usual for the local authority to require the submission of drawings and other information in accordance with its own regulations. Commonly, it will also require compliance with its Rules for Management, which govern the management and maintenance of the premises.

At present there are no nationally recognised technical regulations available to guide applicants for entertainment licences. It is often the case that each local authority has produced its own guidance documentation resulting in considerable variation in the standards applied between (often neighbouring) authorities.

Typically, local authorities will impose standards covering:

- the site of the premises;
- construction and fire separation;
- requirements for boiler and transformer chambers, electrical apparatus, gas metering, etc.;
- requirements for internal car parks;
- detailed provisions covering the building's services;
- premises containing a stage;
- premises used for film exhibitions;
- detailed requirements for fire safety in small premises where permanent provision for a closely seated audience is not made.

Many of these provisions are, of course, covered by other legislation (Building Regulations, Workplace Regulations, etc.). Even so, local authorities are able to apply their own regulations in addition to these other statutory controls and there is always the possibility of conflict between competing standards.

In an attempt to address these inherent difficulties, the London District Surveyors Association (LDSA) has been working for a number of years on the Model Technical Regulations for Places of Public Entertainment. These were in existence for many years under the former Greater London Council and on abolition of the GLC they were adopted by most of the new London Boroughs (sometimes with minor variations). They have also been adopted by a number of local authorities in other parts of England and Wales.

With the advent of the Building Regulations in 1985 the LDSA set up a working party to try to harmonise the Model Technical Regulations with the Building Regulations. This has developed to the extent that a new set of harmonised Model Technical Regulations has been going through a process of consultation, with a strong recommendation that it be adopted by all local authorities on a national basis.

Local authorities are empowered to enforce the provisions of the Act and this includes the powers of entry when an event is taking place or is about to take place. These powers also extend to both the police and the fire officer. Additionally, the fire officer is empowered to enter premises at any time provided he gives at least 24 hours' prior notice.

Persons aggrieved by the refusal of a local authority to grant a licence may appeal to a magistrates' court.

The provisions of the Local Government (Miscellaneous Provisions) Act do not apply in London; however very similar provisions exist by virtue of the London Government Act 1963. The main difference is that the period of notice for a licence application is 21 days.

3.5.4 *The Theatres Act 1968*[11]

The licensing of premises for the public performance of plays is governed by section 12 of the Act. Application (together with the appropriate fee) must be made to the local authority and to the chief officer of police at least 21 days before the intended use is to take place. Local regulations will determine the nature of the application and the information needs of the local authority and can also include a provision for notification of the fire authority. The Model Technical Regulations for Places of Entertainment (see section 3.5.3) include rules for theatres.

Interestingly, theatres are the only premises to be licensed by a local authority, that have the right to sell alcohol without having first to obtain a justices' licence. Additionally, the issue of a theatre licence exempts the proprietor of the premises from any regulations relating to the provision of music or dancing when it forms part of a public performance of the play. This includes incidental music before, during or after the play, provided that it takes less than a quarter of the time of the performance.

The local authority may impose such conditions as it deems appropriate including restriction on, or prohibition of, the sale of alcohol. Fee levels are determined by the local authority and they have the power to mitigate the fee if the performance is for educational or charitable purposes.

3.5.5 *The Cinemas Act 1985*[12]

It is an offence under this Act to give film exhibitions without a licence from the local authority. Additionally, there are special arrangements that control the showing of films to children.

Application for a cinema licence (together with the appropriate fee) must be made to the local authority, the chief officer of police and the fire authority at least 28 days before it is required to come into operation. The local authority may impose conditions and the licence will remain in force for up to one year.

Local authorities are empowered to enforce the provisions of the Act and this includes the powers of entry for licensed premises. These powers also extend to both the police and the fire officer, although the fire officer must give at least 24 hours' prior notice.

Certain premises are exempt from the need to have a licence. Principally, this includes a showing in a private dwellinghouse which is not for private gain, or is for demonstration, advertisement, educational or instructional purposes, and to which the public are not admitted. However, commercial operators who use videos (e.g. night-clubs) may need a cinema licence even though fixed theatre-style seating is not provided. Additionally, the issue of a cinema licence exempts the proprietor of the premises from any regulations relating to the provision of music or dancing when it forms part of a public film exhibition.

Local regulations will determine the nature of the application and the information needs of the local authority. The Model Technical Regulations for Places of Entertainment include rules for cinemas.

3.5.6 *The Explosives Act 1875*[13]

The manufacture and storage of explosives are regulated under the Explosives Act 1875 (as amended). All premises where explosives are manufactured or stored in large quantities must be licensed by the HSE. Smaller stores must either have a licence from the local authority or be registered with it. The HSE and the local authority may impose conditions with the grant of the licence and these normally contain fire safety measures.

3.5.7 *The Caravan Sites and Control of Development Act 1960*[14]

This Act requires that, with certain exceptions listed in Schedule 1 to the Act, no occupier of land shall cause or permit any part of the land to be used as a caravan site unless he is the holder of a site licence issued by the local authority.

The local authority may not refuse a site licence once planning permission has been granted but it is empowered to impose reasonable conditions governing:

- the number and type of caravans and their position on the site;
- the measures to be taken for preventing, detecting and fighting fire;

- the provision and maintenance of adequate sanitary facilities;
- the planting and replanting of the site with trees and bushes.

Any person aggrieved by the imposition of conditions may appeal to the magistrates' court within 28 days of the issue of the licence.

Guidance on the design of caravan sites has been issued by the Secretary of State in the following:

- DoE Circular 119/77[15];
- DoE Circular 22/83[16].

3.5.8 Animal establishments

Premises used on a commercial basis for the keeping of animals are controlled by a licensing system operated by local authorities. When deciding to grant a licence, the local authority must have regard to the protection of the animals in case of fire or in an emergency, in addition to the normal requirements for animal welfare. Full details are given below of the Animal Boarding Establishments Act 1963; the other Acts listed follow a similar pattern.

The Animal Boarding Establishments Act 1963[17]

Any person who keeps a commercial establishment for the boarding of dogs or cats must apply to the local authority for a licence, under the provisions of this Act. The local authority, when considering whether to grant a licence, must have regard to:

- the steps to be taken for the protection of the animals in case of fire or other emergency;
- the construction, size, temperature, lighting, ventilation and cleanliness of the accommodation;
- the number of occupants and exercise facilities provided;
- the welfare facilities provided, such as food, drink and bedding;
- arrangements made for exercising and visiting the animals;
- arrangements made to prevent and control the spread of infectious or contagious diseases among the animals, including the provision of adequate isolation facilities.

The owner is obliged to keep a register of all animals taken into the establishment, which must be available for inspection by the local authority or a veterinary surgeon appointed by the local authority (LA).

Licences must be renewed on an annual basis and the local authority may impose such conditions on granting the licence, as it deems appropriate.

The local authority or a duly authorised veterinary surgeon may enter the premises at any reasonable time to confirm that the provisions of the Act are not being contravened.

Similar provisions apply in relation to the following Acts:

- the Breeding of Dogs Act 1973[18];
- the Pet Animals Act 1951[19];
- the Riding Establishments Act 1964[20] as amended by the Riding Establishments Act 1970[21].

Additionally, the Zoo Licensing Act 1981[22], contains provisions for the fire authority (and certain other interested parties) to comment on any application made to the local authority for a licence to operate a zoo. The local authority must take into account any comments received before granting the licence.

3.6 Registration

3.6.1 Introduction

Control by means of registration tends to be used in those cases where members of the community are thought to need extra protection because they are particularly vulnerable. Therefore the provisions apply, for example, to children, older people and those who are suffering from long-term or mental illness who are in need of residential care. In some cases, it is the premises themselves that are registered for the particular use, and in other cases the person running the establishment will be required to register.

3.6.2 The Children Act 1989[23]

It has long been recognised that children may be particularly vulnerable to abuse when 'cared' for outside the family home. This applies equally whether they are living permanently in an institution or are temporarily placed with a child minder or in a day care centre. The Children Act 1989 contains a number of provisions which deal with the registration of premises used for the residential and day care accommodation of children and/or the individuals who run such premises.

Parts VII and VIII of the Act require that no child shall be cared for and provided with accommodation in a children's home unless the home is registered with the relevant local authority. This applies equally to homes run by voluntary organisations and other homes, although local authority homes are excluded from the need to register. The Act details the duties of the organisations that run the homes and sets out the powers of the local authority regarding supervision.

From a practical viewpoint the Secretary of State may make regulations that set out the standards which must be achieved in such homes. These may relate to the everyday running of the homes as well as to accommodation, staffing and equipment and health and safety standards. Therefore, the local authority should be consulted if it is intended to provide a home either by new work or conversion.

Part X of the Act requires the local authority to keep a register of people who act as child minders on their own domestic premises or who provide day care for children under the age of eight years on other than domestic premises. The local authority is permitted to refuse to register a person if it considers them unfit to look after young children, or if it considers that the premises are unsuitable by virtue of their condition, construction, situation or size.

The local authority (the Social Services Department of the relevant County Council) is permitted to impose such reasonable requirements on the person applying for registration as it deems appropriate. These can include restrictions on the number of children to be cared for, and can relate to the safety (including safety in the event of fire) and condition of the premises and any equipment used on the premises. Additionally, the Secretary of State may make regulations that set out the standards which must be achieved in such premises. Therefore, the relevant local authority should be consulted if it is intended to provide child care premises by either new work or conversion since there are no uniform national standards.

At present, it is common for local authorities to apply standards from the guide to the Children Act[24]. Additionally, the Good Practice Guide issued by the National Children's Bureau[25] may also be recommended.

The local authority has powers to inspect premises at any reasonable time and it may cancel the registration under certain defined circumstances (see the Children Act 1989, section 74).

During 2001 the introduction of the provisions of the Care Standards Act 2000[26] will make significant changes to local councils' responsibilities for child care for children under eight years of age. The Act will remove from councils in England the responsibility for regulating child-minders and day care providers. This responsibility will be transferred to OFSTED (the Office for Standards in Education), a central government department. OFSTED will carry out regulation on a national basis, regulating child care against national standards which will be set by the Department for Education and Skills.

3.6.3 The Registered Homes Act 1984[27]

This Act is concerned with the registration of residential care homes (Part I of the Act), and nursing homes and mental nursing homes (Part II of the Act). Residential care homes provide residential accommodation with both board and personal care for persons who need that care by reason of old age, disablement, past or present dependence on alcohol or drugs, or past or present mental disorder. The Act applies to all such places where care is provided for four or more people.

Anyone intending to carry on a residential care home must apply to the social services department of the local authority (the registration authority) for registration. The registration authority may impose conditions for regulating the number and categories of people who may be accommodated (age, gender, etc.).

The Residential Care Homes Regulations 1984[28] govern the conduct of residential care homes and provide details of the registration procedures and fees required by the registration authority. They also deal with the day-to-day running of the home and specify (in section 10) the facilities and services that should be provided:

- the need for adequate fire safety arrangements;
- provision of suitable welfare facilities (wash-basins, water closets, baths, etc);
- facilities for disabled people;
- the need to keep the home in good repair and structural condition, and reasonably decorated;
- the need to provide wholesome and nutritious food, laundry facilities and medical care.

The person registered is required to consult the fire authority on the need for fire precautions in the home.

At present, local authorities commonly recommend the use of the following guidance publications for residential care homes:

- *Home Life – a code of practice for residential care*[29];
- *Home from Home – Creating a Home from Home*[30].

Nursing homes are premises used for the reception of, and the provision of nursing for, persons suffering from any sickness, injury or infirmity. The definition includes maternity homes and private hospitals.

Mental nursing homes are premises used for the reception of, and the provision of nursing or other medical treatment for, one or more mentally disordered patients.

The Nursing Homes and Mental Nursing Homes Regulations 1984[31] govern the conduct of nursing homes or mental nursing homes and provide details of the registration procedures and fees, deal with the day-to-day running of the home and specify (in section 12) the facilities and services that should be provided. These are similar to those listed above for residential care homes.

Anyone intending to carry on a nursing home or mental nursing home must apply to the Secretary of State, in writing, for registration, and must send the application to the health authority for the area in which the home is situated. When making an application for registration the applicant is required to supply particulars specified in Schedule 2 of the Nursing Homes and Mental Nursing Homes Regulations 1984, including details of any comments made by the fire authority in relation to the home, as the Secretary of State may reasonably require. This would imply that there is a need to consult the fire authority before applying for registration although this is not specifically required by the Regulations.

When satisfied that the applicant is a fit person to run a nursing home or mental nursing home and that certain other conditions are met regarding its organisation, situation, construction, state of repair, staffing and equipment, etc., the Secretary of State will register the applicant in respect of the home named in the application and will issue a certificate of registration.

The Government White Paper *Modernising Social Services* was published on 13 September 1999 and set out the Government's commitment to improving protection for vulnerable people through new inspection systems and stronger safeguards. The White Paper sets out the Government's commitment to introduce a greater degree of consistency in standards for all regulated services, within the development of a new independent regulatory structure. It proposes to establish Commissions for Care Standards (CCSs) at regional level to carry out the regulation of care services. These will be based on the boundaries of the NHS and Social Care Regions, so that there will be eight Commissions in England.

The CCSs will have responsibility for regulating the following services:

- residential care homes for adults;
- nursing homes;
- children's homes;
- domiciliary care providers;
- independent fostering agencies;
- residential family centres;
- boarding schools;
- adoption agencies.

Regulation will apply, as now, to residential care homes and nursing homes registered under the Registered Homes Act 1984. In addition, all residential care homes owned by local authorities will be required to register and will be subject to inspection and enforcement procedures in the same way as voluntary and private care homes.

National standards will be developed and introduced for each of the services to be regulated by the Commissions for Care Standards. A comprehensive set of standards for all regulated services is unlikely to be completed until April 2002, which is the earliest possible date for operation of the CCSs.

3.7 The Building Act and means of escape in existing buildings – additional provisions

3.7.1 Introduction

The following provisions tend to be of academic interest only since they have been largely replaced by more recent legislation. However, they remain on the statute book and illustrate the piecemeal nature of the development of means of escape legislation.

3.7.2 Means of escape in certain high buildings

Section 72 of the Building Act 1984 (Means of escape in certain high buildings) is fully described in Chapter 2. It should be noted that this section applies equally to certain categories of existing *or* proposed buildings which have more than two storeys and contain floors exceeding 20 ft (6.1 m) above the surface of the street or ground on any side of the building. However, for existing buildings, the local authority cannot apply the provisions of section 72 to a building which has a valid fire certificate provided under the Fire Precautions Act 1971 or is a workplace subject to the provisions of the Fire Precautions (Workplace) Regulations 1997.

3.7.3 Means of escape in places of public resort

Section 71 of the Building Act 1984 enables a local authority, after consultation with the fire authority, to require such means of ingress and egress, passages or gangways as they deem satisfactory in the following building types:

- theatre, hall or other building used as a place of public resort;
- restaurant, shop, store or warehouse which employs more than 20 people and to which members of the public are admitted;

- club licensed to serve intoxicating liquor (i.e. registered under the Licensing Act 1964);
- schools which are not exempt from the Building Regulations; *and*
- church, chapel or other place of public worship, but this does not include a private house to which members of the public might be admitted only occasionally. Also excluded from control are any churches and chapels which were so used before 1 October 1937 (i.e. the date of commencement of the Public Health Act 1936[32]).

The satisfactory nature of the means of ingress and egress and the passages and gangways is judged by reference to the purpose for which the building is used and the number of people who are likely to be present at any one time. Furthermore, these must be kept clear and unobstructed when people are assembled in the building.

Finally, it should be noted that the local authority cannot apply the provisions of section 71 to a building which has a valid fire certificate provided under the Fire Precautions Act 1971 or is one which is a workplace subject to the provisions of the Fire Precautions (Workplace) Regulations 1997.

3.8 The London Building Acts

3.8.1 Means of escape in old buildings

Section 35 of the London Building Acts (Amendment) Act 1939 enables local authorities to require means of escape to be provided in certain existing buildings:

- any single family dwelling-house with a storey at a height greater than 42 ft (12.8 m) or which is occupied by more than 20 people;
- any building where more than 10 people are employed to work above the first storey or above 20 ft (6.1 m);
- any flat, inn, hotel, boarding house, hospital, nursing home, boarding school, children's home or other institution, which exceeds two storeys in height and contains a storey which is higher than 20 ft (6.1 m);
- any restaurant, shop, store or warehouse with sleeping accommodation above the ground floor and which exceeds two storeys in height and contains a storey which is higher than 20 ft (6.1 m);
- any place of assembly with a floor area exceeding 500 ft^2 (46 m^2).

For a full description of the sections of the London Building Acts which apply fire safety requirements to buildings, see Chapter 2, section 2.9.

3.9 Houses in multiple occupation

Under the Housing Act 1985 local authorities have the power to require means of escape in case of fire in houses which are occupied by persons not forming a single household (known as houses in multiple occupation or HMOs). In the case of HMOs with at least three storeys and a floor area (all storeys combined) of at least 500 m^2, LAs have a duty to ensure that means of escape are adequate. Before exercising their power or performing their duty in respect of means of escape from fire in an HMO, LAs must consult the fire authority.

Under section 352 of the Housing Act 1985 (as amended) a local housing authority may serve a notice requiring the execution of works to make an HMO fit for the number of occupants. Guidance on the standards expected is given in Department of the Environment Circular 12/92 (in England) or Welsh Office Circular 25/92 (in Wales). The guidance is not mandatory but LAs are expected to have regard to it when considering the exercise of their powers under section 352. A certain amount of flexibility is expected and LAs are reminded in the Circulars that they should try to attain achievable standards since excessively high standards may deter landlords from making accommodation available at all.

An outline of the main provisions of Circular 12/92 is given in Chapter 7.

3.10 References

1 Fire Precautions (Factories, Offices, Shops and Railway Premises) Order 1989 (SI 1989/76).
2 Code of Practice for Fire Precautions in Factories, Offices, Shops and Railway Premises not required to have a Fire Certificate.
3 The Fire Certificates (Special Premises) Regulations 1976 (SI 1976/2003).
4 The Safety of Sports Grounds Act 1975, c.52.
5 The Fire Safety and Safety of Places of Sport Act 1987, c.27.
6 The Safety of Sports Grounds (Accommodation of Spectators) Order 1996 (SI 1996/499).
7 The Safety of Sports Grounds Regulations 1987 (SI 1987/1941).
8 *Guide to Safety at Sports Grounds* (4th edn), Department of National Heritage, 1997.
9 The Licensing Act 1964, c.26.
10 The Local Government (Miscellaneous Provisions) Act 1982, c.30.
11 The Theatres Act 1968, c.54.
12 The Cinemas Act 1985, c.13.
13 The Explosives Act 1875, c.17.
14 The Caravan Sites and Control of Development Act 1960, c.62.
15 DoE Circular 119/77: Model Standards for Caravan Sites (permanent residential caravan sites and holiday caravan sites).

16 DoE Circular 22/83: Model Standards for Caravan Sites (touring caravan sites).
17 The Animal Boarding Establishments Act 1963, c.43.
18 The Breeding of Dogs Act 1973, c.60.
19 The Pet Animals Act 1951, c.35.
20 The Riding Establishments Act 1964, c.70.
21 The Riding Establishments Act 1970, c.32.
22 The Zoo Licensing Act 1981, c.37.
23 The Children Act 1989, c.41.
24 The Children Act 1989, Volume 2, Guidance and Regulations, Family Support, Day Care and Educational Provision for Young Children.
25 *Young Children in Group Day Care, Guidelines for Good Practice.* National Children's Bureau, 1994.
26 Care Standards Act 2000, c.14.
27 The Registered Homes Act 1984, c.23.
28 The Residential Care Homes Regulations 1984 (SI 1984/1345).
29 *Home Life – a code of practice for residential care*; report of a working party sponsored by the Department of Health, 1984.
30 *Home from Home – Creating a Home from Home*; a Guide to Standards; report compiled by the Residential Forum, 1996.
31 The Nursing Homes and Mental Nursing Homes Regulations 1984 (SI 1984/1578).
32 Public Health Act 1936, c.49.

Chapter 4
Means of Escape – General Principles

4.1 Introduction

In Chapter 2 it was shown that the principal legal controls over means of escape in new and altered buildings are through the provisions of the Building Act 1984 and the Building Regulations 2000.

The legal requirement for means of escape is contained in Schedule 1 to the 2000 Regulations. Paragraph B1 reads:

> 'The building shall be designed and constructed so that there are appropriate provisions for the early warning of fire, and appropriate means of escape in case of fire from the building to a place of safety outside the building capable of being safely and effectively used at all material times.'

This is, of course, a functional requirement, which merely states *what* is required without giving details of *how* it may be satisfied.

For smaller, simpler types of buildings, guidance on how to satisfy the functional requirements referred to in paragraph B1 above, may be found in Approved Document B (AD B). This is published by the Department for Transport, Local Government and the Regions (DTLR) and is one of 14 Approved Documents corresponding to the 13 Parts of Schedule 1 to the Building Regulations plus one other which deals with materials and workmanship and gives practical guidance on the requirements of Regulation 7. (Full information regarding the Approved Documents and their legal status is given in Chapter 2.)

AD B sets out the following basic principles whereby the mandatory requirements of paragraph B1 may be met:

- escape routes should be provided which are suitably located and of sufficient number and size to enable the occupants to escape to a place of safety in the event of fire;

- escape routes should be adequately lit and suitably signed;
- appropriate facilities should be provided to limit the ingress of smoke to the escape route or to restrict the fire and remove smoke;
- where necessary, escape routes should be enclosed so that they are sufficiently protected from the effects of fire;

all to an extent necessary depending on the size, height and use of the building.

Additionally, there must, of course, be sufficient means for giving early warning of fire for the building's occupants.

This chapter sets out general principles for the provision of means of warning and escape which apply to all building types.

4.2 Building use and means of warning and escape

The fire hazard presented by a building will, to a large extent, depend on the use to which the building is put. Many of the provisions contained in AD B, including those covering means of warning and escape, are directly related to these use classifications. These are termed 'purpose groups' (PG) in the Approved Document.

The purpose groups are set out in Table D1 (Classification of purpose groups) of Appendix D of AD B. The seven purpose groups are as follows:

1 *Residential (dwellings)* – this includes parts of a dwelling used by the occupant in a professional or business capacity (such as a surgery, consulting room, office or other accommodation), not exceeding 50 m^2 in total. This group is further sub-divided into:

 1(a) flat or maisonette;
 1(b) dwellinghouse (with a habitable storey more than 4.5 m above ground level);
 1(c) dwellinghouse (no habitable storey above 4.5 m from ground level); 1(c) also includes any detached garage or open carport not exceeding 40 m^2 in area, or a detached building consisting of a garage and open carport, neither of which exceeds 40 m^2 in area, irrespective of whether or not they are associated with a dwelling.

2(a) *Residential (institutional)* – includes a hospital, home, school or other similar establishment. The premises will be used as living (and sleeping) accommodation for, or for the treatment, care or maintenance of:

- persons suffering from disabilities due to illness or old age or other physical or mental incapacity;

- children under the age of five years;
- persons in a place of lawful detention.

2(b) *Residential (other)* – includes a hotel, boarding house, residential college, hall of residence, hostel, and any other residential purpose not described above.

3 *Office* – includes offices or premises used for the purpose of:

- administration;
- clerical work (including writing, book-keeping, sorting papers, filing, typing, duplicating, machine calculating, drawing and the editorial preparation of matter for publication, police and fire service work);
- handling money (including banking and building society work);
- communications (including postal, telegraph and radio communications), radio, television, film, audio or video recording or performance (not open to the public) and their control.

4 *Shop and commercial* – shops or premises used for a retail trade or business including:

- the sale to members of the public of food or drink for immediate consumption;
- retail by auction, self-selection and over-the-counter wholesale trading;
- the business of lending books or periodicals for gain;
- the business of a barber or hairdresser;
- premises to which the public is invited to deliver or collect goods in connection with their hire, repair or other treatment, or (except in the case of repair of motor vehicles) where they themselves may carry out such repairs or other treatments.

5 *Assembly and recreation* – places of assembly, entertainment or recreation including:

- broadcasting, recording and film studios open to the public;
- bingo halls, casinos, dance halls;
- entertainment, conference, exhibition and leisure centres;
- funfairs and amusement arcades;
- museums and art galleries;
- non-residential clubs;
- theatres, cinemas and concert halls;
- educational establishments;

- dancing schools, gymnasia, swimming pool buildings, riding schools, skating rinks, sports pavilions, sports stadia;
- law courts;
- churches and other buildings of worship;
- crematoria;
- libraries open to the public;
- non-residential day centres, clinics, health centres and surgeries;
- passenger stations and termini for air, rail, road or sea travel;
- public toilets;
- zoos and menageries.

6 *Industrial* – factories and other premises used for:

- manufacturing, altering, repairing, cleaning, washing, breaking-up, adapting or processing any article;
- generating power;
- slaughtering livestock.

7 *Storage and other non-residential* – this group is further sub-divided into:

7(a) place for the storage or deposit of goods or materials (other than described under 7(b)) and any building not within any of the purpose groups 1 to 6; *and*

7(b) car parks designed to admit and accommodate only cars, motorcycles and passenger or light goods vehicles weighing no more than 2500 kg gross.

Normally the purpose group is applied to the whole building, or (where a building is compartmented) to a compartment in the building, by reference to the main use of the building or compartment. Parts of the building put to different uses can be treated as ancillary to the main use, and therefore as though they are in the same use. However, a different use in the same building is not regarded as ancillary, and is therefore treated as belonging to a purpose group in its own right:

- where the ancillary use is a flat or maisonette;
- where the building or compartment exceeds 280 m^2 in area and the ancillary use exceeds one-fifth of the total floor area of the building or compartment;
- where the building is a shop or commercial building or compartment of purpose group 4 and contains a storage area which exceeds one-third of the total floor area of the building or compartment and the building or compartment is more than 280 m^2 in area.

Where a building contains different main uses which are not ancillary to one another, each use should be considered as belonging to a purpose group in its own right.

Some large buildings, such as shopping complexes, may involve complicated mixes of purpose groups. In these cases special precautions may need to be taken to reduce any additional risks caused by the interaction of the different purpose groups.

4.3 Management of the building and the means of escape

The guidance in AD B is based on the assumption that subject premises will be properly managed. Building owners who do not take proper management responsibility may be subject to prosecution under the Fire Precautions Act or the Health and Safety at Work etc. Act and this can result in prohibition of the use of the premises.

Furthermore, the Fire Precautions (Workplace) Regulations impose a statutory obligation on the employer in charge of a workplace to undertake an assessment of the fire risk. The regulations require that the number, distribution and dimensions of emergency routes and exits shall be adequate having regard to the use, equipment and dimensions of the workplace and the maximum number of persons that may be present at one time. Fire management issues in workplaces (including the Fire Precautions (Workplace) Regulations) are fully discussed in Chapter 11.

It is extremely important to realise that there is a causal connection between the design of the means of warning and escape in a building and the eventual management of that means of warning and escape. Therefore, the escape strategy to be adopted in a particular building will be based on one of the following:

- *Simultaneous evacuation* – all the occupants are expected to leave the building at the same time.
- *Phased evacuation* – only the storeys most affected by the fire (e.g. the floor of origin and the floor above it) are evacuated immediately. Subsequently, two floors at a time are evacuated if the need arises.
- *Progressive horizontal evacuation* – is the concept which is usually adopted in the in-patient parts of hospitals and similar health care premises where total evacuation of the building is inappropriate. In-patients are evacuated, in the event of fire, to adjoining compartments or sub-divisions of compartments, the object being to provide a place of relative safety within a short distance. If necessary, further evacuation can be made from these safe places but under less pressure of time.

Therefore, in buildings other than dwellings, selection of the appropriate fire detection and alarm system will depend on the means of escape strategy adopted. For example, in residential accommodation (where the occupants sleep on the premises) the threat posed by a fire will be much greater than in premises where the occupants are fully alert. In these circumstances an escape strategy based on simultaneous evacuation will mean that all fire sounders will operate almost instantaneously once a manual call point or fire detector has been activated. If, however, the escape strategy is based on phased evacuation, a staged alarm system may be more appropriate. Two or more stages of alarm may be given within a particular area corresponding to 'alert' and 'evacuate' signals.

At the very least the strategy adopted will have implications for the design of fire alarm systems, stairs and exits in many buildings and the fire resistance of partition walls and doors in health care premises.

4.4 Means of giving warning

Clearly, there must be sufficient means for giving early warning of fire for persons in buildings. This does not mean that all buildings must be fitted with automatic fire detection and alarm systems; this will only be necessary if the risk to life warrants it.

4.4.1 Fire detection and alarm in non-residential buildings

All buildings should have arrangements for detecting fire and in most buildings this will be done directly by people through observation or smell. In many small buildings where there is no sleeping risk this may be all that is needed. Similarly, the means of raising the alarm may be simple in such buildings and where the occupants are in sight and hearing of each other a shouted warning may be sufficient. Clearly, it is necessary to assess the risk in each set of circumstances and decide standards on a case-by-case basis. Risk analysis in workplaces is considered in Chapter 9.

The risk analysis will consider the likelihood of a fire occurring and the degree to which the alarm can be heard by all the occupants. Therefore any of the following may need to be incorporated in the building:

- manually operated sounders (e.g. rotary gongs or handbells);
- simple manual callpoints combined with bell, battery and charger;
- electrically operated fire warning system with manual callpoints sited adjacent to exit doors, combined with sufficient sounders to ensure that the

alarm can be heard throughout the building. The system should comply with BS 5839 : Part 1[1] (this is described more fully in Chapter 6) and the call points with BS 5839 : Part 2[2].

Automatic fire detection systems involve a sensor network plus associated control and indicating equipment. Sensors may detect heat, smoke or radiation and it is usual for the control and indicating equipment to operate a fire alarm system. It may also perform other signalling or control functions, such as the operation of an automatic smoke control system.

Automatic fire detection systems are not normally needed in non-residential occupancies; however it may be necessary to install a fire detection system in the following circumstances:

- to compensate for the fact that it has not been possible to follow all the guidance in AD B;
- where it is necessary as part of a fire protection operating system, such as a pressurised staircase or automatic door release mechanism;
- where a fire could occur unseen in an unoccupied or rarely visited part of a building and could prejudice the means of escape from the occupied parts.

Where a building is designed for phased evacuation (see section 4.3), it should be fitted with an appropriate fire warning system conforming to at least the L3 standard given in BS 5839 : Part 1 (see Chapter 6). Additionally, an internal speech communication system (telephone, intercom, etc.) should be provided so that conversation is possible between a fire warden at every floor level and a control point at the fire service access level.

In certain premises where large numbers of the public are present (e.g. large shops and places of assembly) it may be undesirable for an initial general alarm to be sounded since this may cause unnecessary confusion. Therefore it is essential in these circumstances that staff are trained to effect pre-planned procedures for safe evacuation. Usually, actuation of the fire alarm system will alert staff by means of discreet sounders or personal pagers first. This will enable them to be prepared and in position should it be necessary for a general evacuation to be initiated by means of sounders or an announcement over the public address system. In all other respects any staff system should conform to BS 5839 : Part 1. Voice alarms should comply with BS 5839 : Part 8[3].

4.4.2 *Fire detection and alarm in residential buildings*

All dwellinghouses (including bungalows) should be protected by some form of automatic detection and alarm system. The scale of the installation will depend on the size of the house. For example, a small bungalow may only need a single mains-operated smoke alarm. However, a large house (i.e. a house in which any storey exceeds 200 m^2 in area) will need to be fitted with a full L2 system as described in BS 5839 : Part 1 (see Chapter 6), if such a house has more than three storeys, including basements. An automatic smoke detection and alarm system, based on linked smoke alarms, should also be installed where it is proposed to convert the roofspace of a one or two-storey house to habitable accommodation.

Flats and maisonettes (including student residential accommodation which is constructed in the same way as a block of flats) will generally follow the same principles as for dwellinghouses, a maisonette being treated as a two-storey house. Therefore, it will normally be appropriate to provide an automatic detection system within each flat with the warning being given within the flat of fire origin. However, common parts of blocks of flats do not need to be fitted with a warning system and there is no need to interconnect the detection and alarm systems in separate flats. Where the risk analysis reveals that a general evacuation of the whole building is essential the alarm system should be provided in accordance with BS 5839 : Part 1.

Occupancies within purpose groups 2(a) Residential (institutional) and 2(b) Residential (other) (see section 4.2) should be provided with automatic fire detection and alarm systems in accordance with BS 5839 : Part 1.

In a sheltered housing scheme with a warden or supervisor, the fire detection equipment should have a connection to a central monitoring point (or central alarm relay station). This will inform the warden that a fire has occurred and enable him to locate the dwelling of origin of the fire. Common parts of the sheltered housing scheme (e.g. communal lounges) do not need to be linked to this central monitoring point and this provision does not apply to sheltered housing within purpose group 2(a) and 2(b) occupancies.

4.5 General requirements for means of escape

There are certain basic principles which govern the design of means of escape in buildings and which apply to all building types. In general the design should be based on an assessment of the risk to the occupants should a fire occur and should take account of:

- the use of the building (and the activities of the users);
- the nature of the building structure;
- the processes undertaken and/or the materials stored in the building;
- the potential fire sources;
- the potential for fire spread throughout the building;
- the standard of fire safety management to be installed.

In assessing the above, judgements regarding the likely level of provision may have to be made when the exact details are unknown.

The following assumptions must be made in order that a safe and economical design may be achieved:

(1) In general, when a fire occurs, the occupants should be able to escape safely, without external assistance or rescue from the fire service or anyone else. Obviously, there are some institutional buildings where it is not practical to expect the occupants to escape unaided, and special arrangements are necessary in these cases. Similar considerations apply to disabled people. Aided escape is also permitted in certain low-rise dwellings.
(2) Fires do not normally break out in two different parts of a building at the same time.
(3) Fires are most likely to occur in the furnishings and fittings of a building or in other items which are not controlled by the Building Regulations.
(4) Fires are less likely to originate in the building structure and accidental fires in circulation spaces, corridors and stairways are unlikely due to the restriction on the use of combustible materials in these areas.
(5) When a fire breaks out the initial hazard is to the immediate area in which it occurs and it is unlikely that a large area will be affected at this stage. When fire spread does occur it is usually along circulation routes.
(6) The primary danger in the early stages of a fire is the production of smoke and noxious gases. These obscure the way to escape routes and exits and are responsible for the most casualties. Therefore, limiting the spread of smoke and fumes is vital in the design of a safe means of escape.
(7) Buildings covered are assumed to be properly managed. Where there is a failure of management responsibility, the building owner or occupier may be prosecuted under the Fire Precautions Act or the Health and Safety at Work etc. Act, which may result in prohibition of the use of the building.

4.5.1 *Escape route criteria*

When a fire occurs it should be possible for people to turn their backs on it and travel away from it either to a final exit or a protected escape route leading to a place of safety. This means that alternative escape routes should be provided in most situations.

The basic criteria governing the design of means of escape are as follows:

- The first part of the escape route will be within the accommodation or circulation areas and will usually be unprotected. It should be of limited length so that people are not exposed to fire and smoke for any length of time. Where the horizontal escape route is protected it should still be of limited length since there is always the risk of premature failure.

- The second part of the escape route will usually be in a protected stairway designed to be virtually 'fire sterile'. Once inside it should be possible to proceed direct to a place of safety without rushing. Therefore, flames, smoke and gases must be excluded from these routes by fire-resisting construction or adequate smoke control measures or by both these methods. This does not preclude the use of unprotected stairs for normal everyday use; however their relative vulnerability to fire situations means that they can only be of limited use for escape purposes.
- The protected stairway should lead directly to a place of safety or it may do this via a protected corridor. The ultimate place of safety is open air clear of the effects of fire; however, in certain large and complex buildings reasonable safety may be provided within the building if suitable planning and protection measures can be included in the design.

4.5.2 *Alternative means of escape and dead-ends*

Ideally, alternative escape routes should be provided from all points in a building, since there is always the possibility that the path of a single escape route may become impassable due to the presence of fire, smoke or fumes. Escape in one direction only (a dead-end) is acceptable under certain conditions depending on:

- the use of the building;
- its associated fire risk;
- its size and height;
- the length of the dead-end;
- the number of people accommodated in the dead-end.

4.5.3 *Unacceptable means of escape*

Certain paths of travel are not acceptable as means of escape, including:

- *Lifts*, unless designed and installed as evacuation lifts for disabled people in the event of fire.
- *Portable or throw-out ladders.*
- *Manipulative apparatus and appliances* such as fold-down ladders and chutes.
- *Escalators.* These should not be counted as additional escape routes due to

the uneven nature of the top and bottom steps; however it is likely that people would use them in the event of a fire. Mechanised walkways could be acceptable if they were properly assessed for capacity as a walking route in the static mode.

4.5.4 *Security*

It is possible that security measures intended to prevent unauthorised access to a building may hinder the entry of the fire services when they need access to fight a fire or rescue trapped occupants. Advice may be sought from architectural liaison officers attached to most police forces so that possible conflicts between security and access may be solved at the design stage.

4.6 References

1 BS 5839: Fire detection and alarm systems for buildings: Part 1: 1988 Code of practice for system design, installation and servicing.
2 BS 5839: Fire detection and alarm system for buildings: Part 2: 1983 Specification for manual call points.
3 BS 5839: Fire detection and alarm systems for buildings: Part 8 1998 Code of practice for the design, installation and servicing of voice alarm systems.

Chapter 5
Means of Escape – Principles in Practice

5.1 Introduction

Chapter 4 sets out general principles for the provision of means of warning and escape which apply to all building types. This chapter takes those general principles and uses the recommendations of sections 1 to 6 of AD B to develop a simple evacuation model for the design of means of escape in other residential buildings (hotels, hostels, halls of residence, small nursing homes and other similar institutional buildings where people sleep on the premises) and smaller, simpler buildings used for other non-residential purposes. More detailed information or guidance which is specific to particular building uses is covered in Chapters 6 and 7.

In more complex buildings (and buildings involving multiple uses) the simple approach offered by AD B may prove inappropriate. In these cases the basic principles can be modified (see Chapter 8) or it may be better to use the different guidance documents mentioned in Chapters 6 and 7 or a fire safety engineering approach (see Chapter 10).

5.2 A strategy for design

The following evacuation model is intended to be used as a guide to the design of means of escape in:

- small nursing homes and other similar institutional buildings in purpose group 2(a) where people sleep on the premises;
- other residential buildings (hotels, hostels, halls of residence, etc. in purpose group 2(b)); *and*
- smaller, simpler buildings in any of the other non-residential purpose groups.

Whilst many of the principles discussed apply equally to dwellings (houses, flats and maisonettes), these are more specifically covered in Chapter 6.

The model is based on the recommendations of sections 2 to 6 of AD B and is illustrated in outline in Fig. 5.1. It should be read in conjunction with the case study in Appendix A, which looks at the design of the means of escape in a simple five-storey office block of rectangular plan (see Fig. A1).

5.2.1 *Step 1 - Assessment of use (purpose group)*

Deciding the purpose group into which the building falls affects the calculation of occupant capacity (especially where the actual population density is not known for certain), and has implications for allowable travel distances and widths of escape stairs in AD B1. The choice of purpose group also affects a number of other parameters in other parts of Approved Document B, such as the recommended minimum period of fire resistance of the building and the permitted limits of unprotected areas in the external walls. When deciding the purpose group of the building the following considerations should be borne in mind:

- Normally the purpose group is applied to the whole building by reference to its main use; however where a building contains compartments used for different purposes, each compartment should be classified separately. In this respect 'compartment' is defined in AD B as 'a building or part of a building, comprising one or more rooms, spaces or storeys, constructed to prevent the spread of fire to or from another part of the same building, or an adjoining building'. A roof space above the top storey of a compartment is included in that compartment.
- Reference to section 4.2 will show that it is acceptable to treat a different use in the same building as ancillary to the main use rather than belonging to a purpose group in its own right, except:
  - where the ancillary use is a flat or maisonette;
  - where the building or compartment exceeds 280 m^2 in area and the ancillary use exceeds one-fifth of the total floor area of the building or compartment;
  - where the building is a shop or commercial building or compartment of purpose group 4 and contains a storage area which exceeds one-third of the total floor area of the building or compartment and the building or compartment is more than 280 m^2 in area.
- Where the building is divided into different occupancies with different uses, each occupancy should be classified separately.

5.2.2 *Step 2 – Calculation of occupant capacity*

In order to design a safe means of escape, it is necessary to assess the number of people who are likely to be present in the different parts of the building (i.e. the occupant capacity). Occupant capacity depends partly on the use (or purpose group) of the building and partly on the use of individual rooms in the building, and has a direct effect on the numbers and widths of:

- the exits from any room, storey or tier; *and*
- escape stairs and final exits from the building.

For some building uses (such as theatres or restaurants) occupant capacity can be calculated by totalling the number of seats and then adding an allowance for staff.
In other buildings (such as speculative office developments, shops and supermarkets), the designer will not be able to do this since he will never be sure how intensively the building will be used. In these circumstances, Table 1 of AD B1 provides floor space factors (expressed in m^2 per person) which will give a value for the occupant capacity when divided into the relevant floor area. The table is based on the type of accommodation contained in the building (i.e. the use of individual rooms) and therefore the following procedure should be adopted to calculate the occupant capacity for the building:

- choose the top floor and inspect each room to determine its use;
- for each room calculate its floor area and divide it by the relevant floor space factor to determine the occupant capacity for the room;
- add together the individual room capacities to obtain the occupant capacity for the floor;
- repeat this process for each floor;
- total the occupants of all floors to obtain the occupant capacity for the building.

Floor space factors based on Table 1 of AD B1 are given in Table 5.1. The following should also be noted when using the floor space factors:

- it is not necessary to calculate occupancy capacities for stair enclosures, lifts or sanitary accommodation, and fixed parts of the building structure may also be excluded from the calculation (although the area taken up by counters and display units, etc. should not be excluded);
- where the descriptions do not completely cover the accommodation it is acceptable to use a value based on a similar use;
- where any part of the building is likely to have multiple uses (for example, a

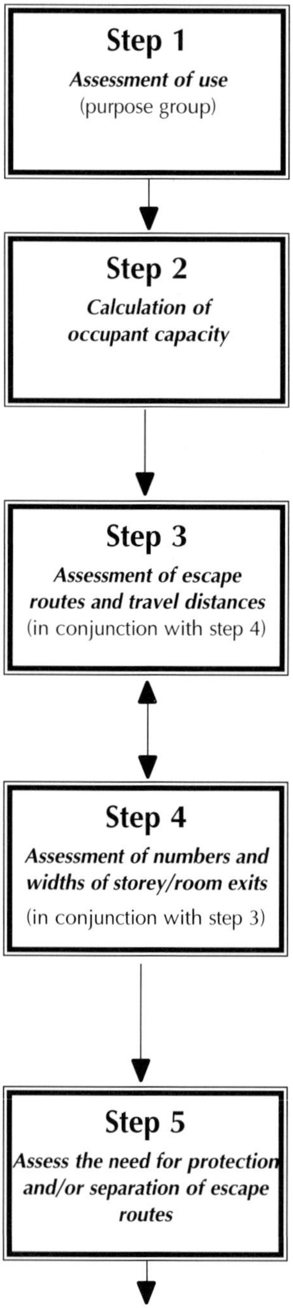

General

Step 1
Assessment of use
(purpose group)

Decide the main use(s) to which the building will be put

Are there any ancillary uses?
Is the building divided into different occupancies/uses?
Is the building divided into different compartments with different uses?

Step 2
Calculation of occupant capacity

Estimate the total number of people who are likely to be in the building

Are the numbers known accurately (such as in a theatre or restaurant)?
Are floor space factors available from actual data taken from similar premises?
If not, use floor space factors from Table 1 of AD B1.
Calculate number of occupants per floor/use/occupancy/compartment.

Horizontal escape

Step 3
Assessment of escape routes and travel distances
(in conjunction with step 4)

Check maximum travel distances from scaled floorplans

Are single and multi-direction paths of travel within the distance limitations given in AD B1 Table 3?
Are assumed alternative paths of travel true alternatives?
Adjust floor layout and position of storey/room exits to accommodate maximum allowable travel distances.

Step 4
Assessment of numbers and widths of storey/room exits
(in conjunction with step 3)

Check that assumed numbers and widths of exits can cater for expected occupant numbers at each level

Are enough exits provided from the room/storey? (See AD B1 Table 4).
Are widths of exits sufficient to take the expected numbers of occupants (including the possibility of discounting)? (See AD B1 Table 5).
Adjust floor layout and numbers/widths of storey/room exits to accommodate expected numbers of occupants.

Step 5
Assess the need for protection and/or separation of escape routes

Establish if any of the identified horizontal escape routes need to be fire protected and/or separated from other parts of the building

Do any dead-end situations exist?
Are there any corridors serving bedrooms?
Are any corridors common to two or more different occupancies?
Do any corridors connect alternative escape routes?
Are any basements present?
Are any inner rooms present?

Fig. 5.1 Means of escape – simple evacuation model.

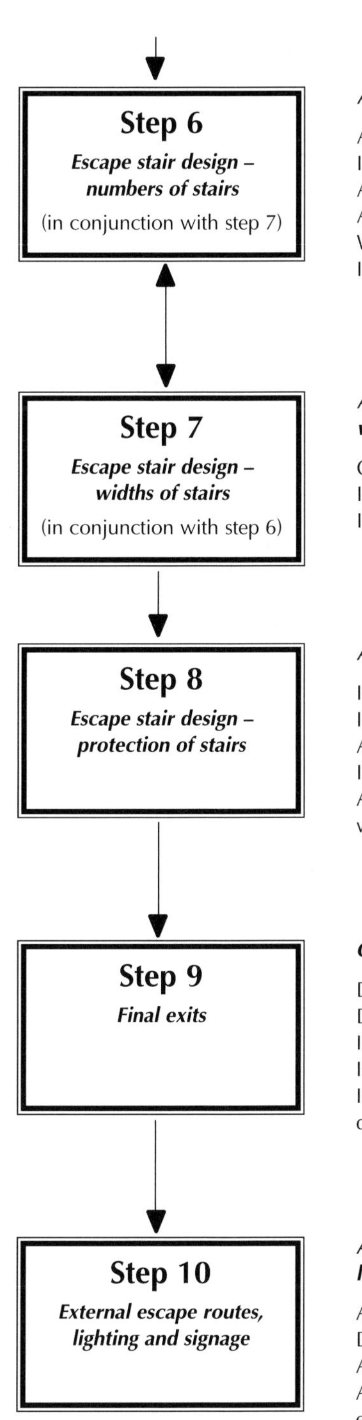

Vertical escape

Assess the number of escape stairs needed

Are sufficient storey exits provided? (See steps 3 and 4).
Is a single stair acceptable?
Are independent stairs required (mixed occupancy buildings)?
Are fire-fighting stairs needed?
Will it be necessary to discount any of the stairs?
Is access needed to basements?

Assess widths of stairs needed and check that assumed numbers/ widths can cater for expected occupant numbers

Check that minimum stair widths are provided. (See AD B1 Table 6).
Is escape based on simultaneous evacuation? (See AD B1 Table 7).
Is escape based on phased evacuation? (See AD B1 Table 8).

Assess the need for protection to the escape stairs

Is an unprotected stairway acceptable?
Is the stairway a protected shaft or a fire-fighting shaft?
Are access lobbies or corridors needed to protect the stairway?
Is the stairway to be used for anything other than access and escape?
Are the external walls of the stairway protected from fire occurring within the building?

Exits from escape routes

Check design of exits from the building/protected stairway

Does the protected stairway discharge directly to a final exit?
Does it discharge by way of a protected exit passageway to a final exit?
Is the position of the final exit clearly apparent to users?
Is width of the final exit adequate for the expected occupant numbers?
Is final exit sited clear of any risk areas so as to ensure safe and rapid dispersal of occupants?

Further design considerations

Assess the design of any external escape routes, the provision of lighting, and escape route signage

Are there any external escape routes?
Do they pass over any flat roofs?
Are all escape routes adequately lit?
Are all escape routes distinctively and conspicuously marked by emergency exit signs?

Table 5.1 Floor space factors.

Floor space factor (m^2 per person)	Type of accommodation/use of room	Notes
0.3	Standing spectator areas. Bars and other similar refreshment areas, without seating	
0.5	Amusement arcade, assembly hall (including a general purpose place of assembly), bingo hall, crush hall, dance floor or hall, venue for pop concert or similar events	
0.7	Concourse, queuing area or shopping mall	See also section 4 of BS 5588 : Part 10: 1991[1] Code of practice for shopping complexes for detailed guidance on the calculation of occupancy in common public areas in shopping complexes.
1.0	Committee room, common room, conference room, dining room, licensed betting office (public area), lounge or bar (other than above), meeting room, reading room, restaurant, staff room or waiting room	In many of these uses the occupants will normally be seated. In such cases occupant capacity may be taken as the number of *fixed* seats provided.
1.5	Exhibition hall or studio (film, radio, television, recording)	
2.0	Skating rink, shop sales area (1)	(1) Shops, such as supermarkets and the main sales areas of department stores; shops for personal services (e.g. hairdressers); shops for the delivery or collection of goods for cleaning, repair or other treatment either by the company or by the public themselves. For other shops see (2) below.
5.0	Art gallery, dormitory, factory production area, museum or workshop	
6.0	Office	
7.0	Kitchen or library, shop sales area (2)	(2) Shops trading in large items such as furniture, floor coverings, cycles, prams, large domestic appliances or other bulky goods, or cash and carry wholesalers. For other shops see (1) above.
8.0	Bedroom or study-bedroom	
10.0	Bed-sitting room, billiards or snooker room or hall	
30.0	Storage and warehousing	
	Car park	Occupant capacity based on two persons per parking space.

multi-purpose hall might be used for a low density activity like gymnastics or a high density use such as a disco) the most onerous floor space factor should be used.

Alternatively, some designers or developers may be able to obtain actual data relating to occupancy numbers, taken from existing premises which are similar to those being designed. Where this data is available it should reflect the average occupant density at a peak trading time of year.

5.2.3 Step 3 – Assessment of escape routes and travel distances

Once an estimate has been made of the occupant capacity of each room, storey or tier in the building (see step 2), it is possible to calculate the number of escape routes and exits needed, which in turn will depend on the distance that has to be travelled from any point in the building to the nearest exit.

Ideally, there should be alternative escape routes provided from every part of the building in order to prevent occupants from becoming trapped by fire or smoke. In a limited range of circumstances single-direction escape routes are permitted and these are discussed in detail for specific purpose groups in Chapters 6 and 7.

AD B1 places limits on the travel distance from any part of a room, tier or storey of a building to a storey exit by reference to the purpose group of that building, and these are contained in Table 3 from AD B1. The substance of Table 3 is summarised in Table 5.2 of this book. It should be read in conjunction with the following comments:

- The dimensions in the table are actual travel distances and are measured along the shortest route taken by a person escaping in the event of a fire.
- Where there is fixed seating or there are other fixed obstructions, the travel distance is measured along the centre line of the seatways or gangways.
- Where the route of travel includes a stair it is measured along the pitch line on the centre line of travel.
- Where the layout of a room or storey is not known at the design stage, the direct distance measured in a straight line should be taken. Direct distances should be taken as two-thirds of the travel distance.
- Once it has been established that *at least one exit* is within the distance limitations given in Table 5.2 the other exits may be further away than the distances given.

Although a choice of escape routes may be provided from a room or storey, it is possible that they may be so located, relative to one another, that a fire might

Table 5.2 Travel distance limitations.

Purpose group	Maximum travel distance (m) in: One direction	Multi-direction	Notes
2(a) Institutional	9	18	In hospitals or other health care premises where the means of escape is being designed using the Department of Health's *Firecode*[2] documents, the relevant travel distances recommended in those documents should be used.
2(b) Other residential			
(i) in bedrooms	9	18	This is the maximum part of the travel distance within the bedroom but includes any associated dressing room, bathroom or sitting room, etc. It is measured to the door onto the protected corridor serving the bedroom or suite.
(ii) in bedroom corridors	9	35	This is the distance from the door onto the protected corridor serving the bedroom or suite to the storey exit.
(iii) elsewhere	18	35	
3 Office	18	45	
4 Shop and commercial	18	45	For shopping malls see BS 5588 : Part 10. This document applies more restrictive provisions to units with only one exit in covered shopping complexes. See also BRE Report[3] (BR 368) *Design methodologies for smoke and heat exhaust ventilation* for guidance on associated smoke control measures.
5 Assembly and recreation			
(i) buildings mainly for disabled people (not schools)	9	18	
(ii) schools	18	45	
(iii) areas with seating in rows	15	32	
(iv) elsewhere	18	45	

Cont.

Table 5.2 Continued.

Purpose group	Use of the premises or part of the premises			Notes
6 7	Industrial and Storage and other non-residential			
	(i) 'normal' fire risk	25	45	For 'normal' fire risk as defined in Home Office *Guide to fire precautions in existing places of work that require a fire certificate: factories, offices, shops and railway premises*[4].
	(ii) 'high' fire risk	12	25	For 'high' fire risk as defined in Home Office *Guide to fire precautions in existing places of work that require a fire certificate: factories, offices, shops and railway premises.*
2–7	Places of special fire hazard	9	18	This is the maximum part of the travel distance within the room or area. The travel distance outside such room or area should comply with the limits for the purpose group as shown above. Places of special fire hazard are: oil-filled transformer and switch gear rooms, boiler rooms, storage space for fuel or other highly flammable substances, and rooms housing fixed internal combustion engines. Plus, in schools: laboratories, technology rooms with open heat sources, kitchens and stores for PE mats or chemicals.
2–7	Plant room or rooftop plant			
	(i) distance within plant room	9	35	
	(ii) escape route not in open air	18	45	Overall travel distance
	(iii) escape route in open air	60	100	Overall travel distance

disable them both. In order to consider them as true alternatives they should be positioned as shown in Fig. 5.2, i.e. the angle which is formed between the exits and any point in the space should be at least 45°. Where this angle cannot be achieved:

- the maximum travel distance for escape in one direction will apply; *or*
- the alternative escape routes should be separated from each other by fire-resisting construction.

The floor plans may need to be adjusted at this stage to take account of the restrictions on travel distance mentioned above. Adjustment may also be necessary in conjunction with step 4 below when decisions are finalised regarding the numbers and widths of exits needed from a room, tier or storey.

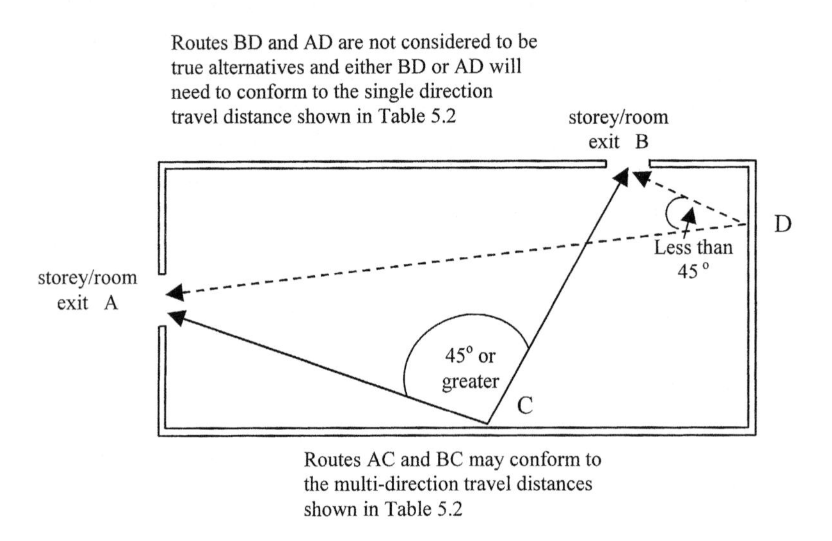

Fig. 5.2 Alternative escape routes.

5.2.4 *Step 4 – Assessment of numbers and widths of storey/room exits*

The number of occupants in a room, tier or storey influences the numbers and widths of the exits and escape routes that need to be provided from that area. Table 4 of AD B1 lists the minimum number(s) of escape routes or exits which should be provided relative to the maximum numbers of occupants:

- up to 60 persons - 1 exit
- 61 to 600 persons - 2 exits
- over 600 persons - 3 exits

Realistically, the figures given above will only serve to define the absolute minimum number of exits for a means of escape. In practical terms the actual number of exits will be determined by travel distances and exit widths. The width of an escape route or exit may be determined by reference to Table 5 of AD B1. The information contained in this table is restructured in Table 5.3 to include relevant data from Approved Document M (Access and facilities for disabled people) since the minimum widths shown in Table 5 of AD B1 may not be adequate for disabled access.

Usually, the narrowest part of an escape route will be at the door openings which form the room or storey exits. These are measured as shown in Fig. 5.3.

Escape route widths are measured at a height of 1500 mm above floor level where the route is defined by walls. Elsewhere, the width will be measured between any fixed obstructions. Stair widths are measured clear between walls and balustrades. Strings may be ignored as may handrails which project less than 100 mm. Additionally, where a stairlift is installed, the guide rail may be ignored; however, the chair or carriage must be capable of being parked where it does not obstruct either the stair or the landing.

Where a storey has two or more exits it is assumed that one of them will be disabled by a fire. Therefore the remaining exits should have sufficient width to take the occupants safely and quickly. This means that the widest exit should be discounted and the remainder should be designed to take the occupants of the storey. Since stairs need to be as wide as the exit leading onto them, this recommendation for exit width may influence the width of the stairways. (Stairways may also need to be discounted and this is discussed in section 5.2.6.)

Except in doorways, all escape routes should have clear headroom of at least 2 m.

5.2.5 Step 5 – Assess the need for protection and/or separation of escape routes

It will be observed from Table 5.2 that where escape is possible in one direction only, the travel distances are much reduced. However, where a storey exit can be reached within these one-directional travel distances, it is not necessary to provide an alternative route except in the case of a room or storey that:

(1) has an occupant capacity exceeding 60 in the case of places of assembly or bars; *or*
(2) has an occupant capacity exceeding 30 if the building is in purpose group 2(a) Residential (institutional);
(3) is used for in-patient care in hospitals.

Table 5.3 Widths of exits and escape routes.

Max. number of persons	Min. width of exit* (mm)	Min. width of escape route* (mm)	Notes
Up to 50	750	750(1)	(1)Does not apply to: • schools where minimum width in corridors is 1050 mm (and 1600 mm in dead-ends) • areas accessible to disabled people where minimum width in corridors is 1200 mm (or 1000 mm where lift access is not provided to the corridor or it is situated in an extension approached through an existing building) • gangways between fixed storage racking in purpose group 4 (Shop and commercial) where minimum width may be 530 mm (but not in public areas) Widths of escape routes and exits less than 1050 mm should not be interpolated
51 to 110	850	850(2)	(2)Does not apply to: • schools where minimum width in corridors is 1050 mm (and 1600 mm in dead-ends) • areas accessible to disabled people where minimum width in corridors is 1200 mm (or 1000 mm where lift access is not provided to the corridor or it is situated in an extension approached through an existing building) Widths of escape routes and exits less than 1050 mm should not be interpolated
111 to 220	1050	1050(3)	(3)Does not apply to: • schools where minimum width in corridor dead-ends is 1600 mm • areas accessible to disabled people where minimum width in corridors is 1200 mm
Over 220	5/person	5/person	This method of calculation should not be used for any opening serving less than 220 persons (e.g. three exits each 850 mm wide will accommodate $3 \times 110 = 330$ people, not the 510 (i.e. $2550 \div 5$) people that $3 \times 850 = 2550$ mm would accommodate.)

* For method of measuring widths of escape routes and exits see section 5.2.4.

Similarly, it is often the case that there will not be alternative escape routes, especially at the beginning of an escape route. A room may have only one exit onto a corridor from where it may be possible to escape in two directions. This is permissible provided that:

- the overall distance from the furthest point in the room to the storey exit complies with the multi-directional travel distance from Table 5.2; *and*

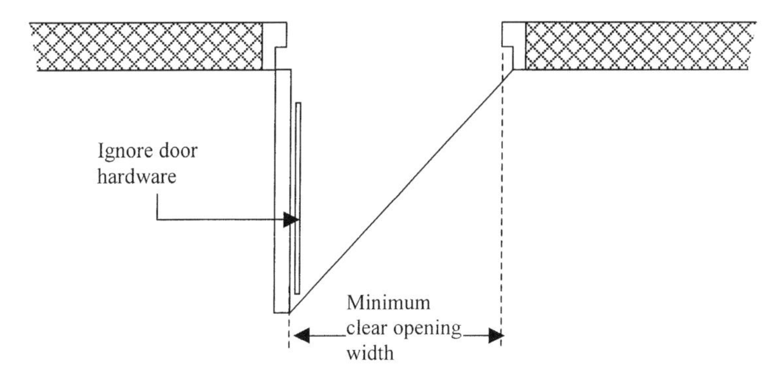

Fig. 5.3 Door width measurement.

- the single direction part of the route (in this case, in the room) complies with the 'one direction' travel distance specified in Table 5.2.

Special rules apply where a dead-end situation exists in an open storey layout as shown in Fig. 5.4:

- XY should be within the one direction travel distance from Table 5.2;
- whichever is the least distance of WXY and ZXY should be within the multi-direction travel distance from Table 5.2;
- angle WXZ should be at least 45° plus 2.5° for each metre travelled in a single direction from Y to X.

Therefore, with the exception of the cases listed above, the following parts of escape routes should be protected with fire-resisting construction:

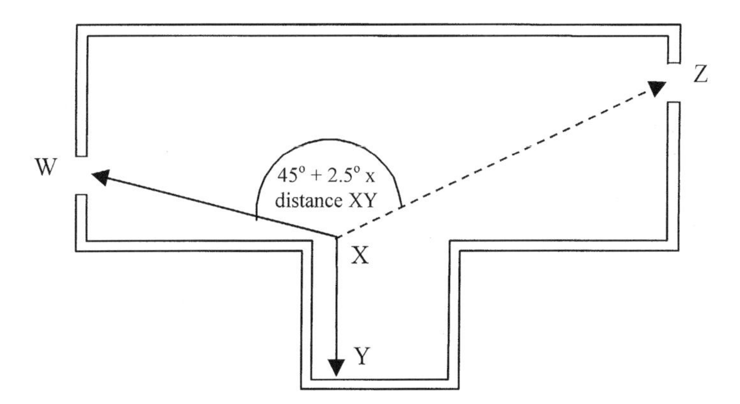

Fig. 5.4 Dead-end situation – open-storey layout.

- every dead-end corridor (although small recesses and extensions less than 2 m long and referred to in figures 10 and 11 of BS 5588 : Part 11[5] may be ignored);
- every corridor serving bedrooms;
- every corridor or circulation space common to two or more 'different occupancies' (i.e. where the premises are split into separate ownerships or tenancies of different organisations). In this case the need for fire protection may be omitted where an automatic fire detection and alarm system is installed throughout the storey. Even so, the means of escape from one occupancy should not pass through any other occupancy.

The way in which a storey layout is planned can have an effect on its means of escape characteristics. For example, whilst it is perfectly acceptable to have open plan floor areas, they offer no impediment to smoke spread but do have the advantage that occupants can become aware of a fire more quickly. On the other hand, the provision of a cellular layout, where the means of escape is enclosed by partitions, means that some defence is provided against smoke spread in the early stages of a fire even though the partitions may have no fire-resistance rating. To maintain the effectiveness of the partitions, they should be carried up to ceiling level (i.e. either to the soffit of the structural floor above or to a suspended ceiling) and room openings should be fitted with doors (which do not need to be fire-resisting).

Corridors which give access to alternative escape routes may become blocked by smoke before all the occupants of a building have escaped and may make both routes impassable. Additionally, the means of escape from any permitted dead-end corridors may be blocked. Therefore, corridors connecting two or more storey exits should be sub-divided by means of self-closing fire doors (and screens, if necessary) if they exceed 12 m in length. The doors (and screens) should be positioned so that:

- they are approximately mid-way between the two storey exits; *and*
- the route is protected from smoke, having regard to any adjacent fire risks and the layout of the corridor.

Unless the escape stairway and its associated corridors are protected by a pressurisation system complying with BS 5588 : Part 4[6], dead-end corridors exceeding 5.5 m in length giving access to a point from which alternative escape routes are available, should also be provided with fire doors so positioned that the dead-end is separated from any corridor which:

- provides two directions of escape; *or*
- continues past one storey exit to another.

Where a building contains a basement, special rules apply to the way in which the staircases are permitted to connect with it (see step 6 below). Generally, any staircase connecting with a basement will need to be separated from it by a protected lobby or there will need to be a protected corridor between the staircase and the basement accommodation.

Special considerations apply where the means of escape from a room (termed an inner room) is only possible by passing through another room (termed an access room). Inner rooms are only acceptable under the following conditions:

- the occupant capacity of the inner room should not exceed 60 (or 30 for institutional buildings in purpose group 2a);
- the inner room should not be a bedroom;
- only one access room should be passed through when escaping from the inner room;
- the inner room should be accessed directly off the access room;
- the maximum travel distance from the furthest point in the inner room to the exit from the access room should not exceed the appropriate limit given in Table 5.2 above;
- the access room should be in the control of the same occupier as the inner room;
- the access room should not be a place of special fire hazard (e.g. a boiler room).

Where these conditions are met the inner room should be designed to conform to one of the following arrangements:

(1) the walls or partitions of the inner room should stop at least 500 mm from the ceiling; *or*

(2) a vision panel, which need not be more than 0.1 m^2 in area, should be situated in the walls or door of the inner room (this is to enable the occupiers to see if a fire has started in the access room); *or*

(3) a suitable automatic fire detection and alarm system should be fitted in the access room which will give warning of fire in that room to the occupiers of the inner room.

5.2.6 *Step 6 – Escape stair design: numbers*

In multi-storey buildings, it is important to ensure that a sufficient number of adequately sized and protected escape stairs are provided. Steps 3 and 4 above deal with the provision of storey exits, and this in turn will dictate the number of escape stairs needed. There are, however, certain additional factors which may need to be considered when choosing the number of escape stairs needed. Examples are given here.

Is it possible for the building to be served by a single escape stair?

Assuming that the building is not excluded from having a single escape route by virtue of steps 3 and 4, it may be served by a single escape stair in the following circumstances:

(1) where it consists of a flat or maisonette and it complies with the recommendations given in section 4 of AD B1 (see Chapter 6);
(2) where it serves a basement which is allowed to have a single horizontal escape route (i.e. storey occupancy not exceeding 60, maximum travel distance within the limits for travel in one direction);
(3) where it serves a building which has no floor more than 11 m above ground which is allowed to have a single horizontal escape route (but for schools see qualifications in Chapter 7, section 7.8);
(4) where it serves what is termed 'small premises' and the recommendations of clause 10 of BS 5588 : Part 11 are followed (see section 5.2.8 and Chapter 7).

Does the building contain a mix of different uses?

In buildings where there is a mix of different uses it may be the case that a fire in an unattended shop or office might have serious consequences for a residential or hotel use in the same building. Therefore, it is important to analyse the risks involved and to consider whether completely separate escape routes should be provided from each different use or whether other effective means of protecting common escape routes can be provided. (See Chapter 6 for examples of the use of common stairs in buildings which contain both dwellings and other uses.)

Are fire-fighting stairs needed?

In large buildings, provisions for fire service access in AD B5 may mean that some of the escape stairs need to serve as fire-fighting stairs. This may affect the total number of escape stairs. Generally, fire-fighting stairs should be provided as shown in Table 5.4.

Will it be necessary to discount any of the stairs?

Where two or more stairways are provided, it is possible that one of the stairs may be inaccessible due to fire or smoke unless special precautions are taken.

Table 5.4 Provision of fire-fighting stairs.

Height (and area if applicable) of floor relative to fire service access level	Use of building	Purpose group	Number and location[a] of fire-fighting shafts containing stairs and lift for:		
			Building not provided with sprinklers	Building provided with sprinklers	
				Largest qualifying floor area (m^2)	Number of shafts
Any basement below 10 m	All	All	One for every 900 m^2 (or part thereof) of floor area		
Two or more basements each more than 900 m^2 in area[b]	All	All		Less than 900	1
Any floor above 7.5 m with storey more than 900 m^2 in area[b]	Shop and commercial Industrial Storage	4 6 7a	One for every 900 m^2 (or part thereof) of floor area	900 to 2000 over 2000	2 2 plus 1 for each additional 1500 m^2 or part thereof
Any floor above 18 m	All	All	One for every 900 m^2 (or part thereof) of floor area of the largest floor above 18 m		

[a] Fire-fighting shafts should be located so that every part of every storey (other than at fire service access level) is within 60 m of the fire main outlet measured on a route suitable for laying a hose. This is reduced to a direct distance of 40 m where the internal layout is not known at the design stage.
[b] In these cases fire-fighting lifts do not have to be provided.

Therefore, it may be necessary to discount each stair in turn in order to check that the remaining stairways are capable of coping with the demand. Discounting is unnecessary if:

(a) the escape stairs are approached through a protected lobby at each floor level (although a lobby is not needed for the top floor for the exception still to apply);
(2) the stairs are protected by a pressurisation smoke control system designed in accordance with BS 5588 : Part 4.

Is access needed to basements?

Basement fires are particularly serious since combustion products tend to rise and find their way into stairways unless other smoke venting measures are taken (see AD B5). Therefore, it is necessary to take additional precautions to prevent a basement fire endangering upper storeys in a building:

- In most buildings with only one escape stair serving the upper storeys, this stair should not continue down to the basement, i.e. the basement should be served by a separate stair. (The only exception to this rule is where the basement forms part of a small single-stair building containing flats or maisonettes – see Chapter 6.)
- In buildings containing more than one escape stair, at least one of the stairs should terminate at ground level and not continue down to the basement. The other stairs may terminate at basement level on condition that the basement accommodation is separated from the stair(s) by a protected lobby or corridor at basement level.

These provisions apply to all buildings, including flats and maisonettes.

5.2.7 Step 7 – Escape stair design: widths

Clearly, the width of escape stairs is related to the number of people that they can carry in an evacuation situation. AD B1 contains a number of provisions which enable the width of stairs to be calculated by reference to:

(1) the number of people who will use them; *and*
(2) their mode of use (i.e. simultaneous or phased evacuation).

Escape stairs should be at least as wide as any exits giving access to them and should not reduce in width as they approach the final exit. Additionally, if the exit route from a stair also picks up occupants of the ground and basement storeys, it may need to be increased in width accordingly (see step 9 below).

Although stairs need to be sufficiently wide for escape purposes, research has shown that people prefer to stay close to a handrail when making a long descent. Therefore the centre of a very wide stairway would be little used and might, in fact, be hazardous. For this reason, AD B1 puts a maximum limit of 1400 mm on the width of a stairway where its vertical extent exceeds 30 m unless it is centrally divided with a handrail. Where the design of the building calls for a stairway that is wider than 1400 mm, it should be at least 1800 mm wide and contain a central handrail. In this case, the stair width on either side of the central handrail will need to be considered separately when assessing stair capacity.

Minimum stair widths can, in the first instance, be assessed using Table 5.5. This is based on Table 6 from AD B1 and is suitable for most simple building

Table 5.5 Minimum widths of escape stairs.

Description of stair	Numbers of people assessed as using stair in emergency[a]	Minimum width of stair (mm)
Escape stairs in any building	Up to 50	800[b]
(but see footnotes for exceptions)	51 to 150	1000
	151 to 220	1100
	Over 220	See note[c]

Notes
[a] For methods of assessing occupancy see section 5.2.2.
[b] This minimum stair width does not apply:
a) In an institutional building unless the stair will only be used by staff
b) In an assembly building unless the area served is less than $100\,m^2$ and/or is not for assembly purposes (e.g. office)
c) To any areas which are accessible to disabled people.
[c] See AD B1 Table 7 (and Formula 5.1) for simultaneous evacuation and AD B1 Table 8 (and Formula 5.2) for phased evacuation.

designs where the maximum number of people served by the stair(s) does not exceed 220.

As Table 5.5 suggests, in multi-storey buildings where the number of occupants exceeds 220 it may be necessary to consider the mode of evacuation and use other methods to calculate stair widths.

Where it is assumed that all the occupants would be evacuated at once, this is termed 'simultaneous evacuation' and this should be the design approach for:

- all stairs which serve basements;
- all stairs which serve buildings with open spatial planning (i.e. where the building is arranged internally so that two or more floors are contained within one undivided volume); *and*
- all stairs which serve Assembly and recreation buildings (PG 5) or Residential (other) buildings (PG 2b).

Using this approach the escape stairs should be wide enough to allow all the floors to be evacuated simultaneously. The calculations take into account the number of people temporarily housed in the stairways during evacuation.

A simple way of assessing the escape stair width is to use Table 7 from AD B1. This covers the capacity of stairs with widths from 1000 mm to 1800 mm for buildings up to 10 storeys high (although the capacity of stairs serving more than 10 storeys can be obtained from Table 7 by using linear extrapolation). In fact, the capacities given in the table for stair widths of 1100 mm and greater are derived from the formula:

$$P = 200w + 50(w - 0.3)(n - 1) \qquad \text{(formula 5.1)}$$

where
P = the number of people who can be served by the stair,
w = the width of the stair in metres, and
n = the number of storeys in the building.

Formula 5.1 can be used for any size of building with no limit being placed on the occupant capacity or number of floors and it is probably advisable to use it for buildings which are larger than those covered by Table 7. It should be noted that separate calculations should be made for stairs serving basements and for those serving upper storeys. Use of formula 5.1 is illustrated in the case study in Appendix A at the end of the book.

The formula is particularly useful in larger buildings which lie outside the range of Table 7. However, it cannot be used for stairs which are narrower than 1100 mm, so for stairs which are allowed to be 1000 mm wide, in buildings up to 10 storeys high, the values have been extracted from Table 7 and are presented in Fig. 5.5.

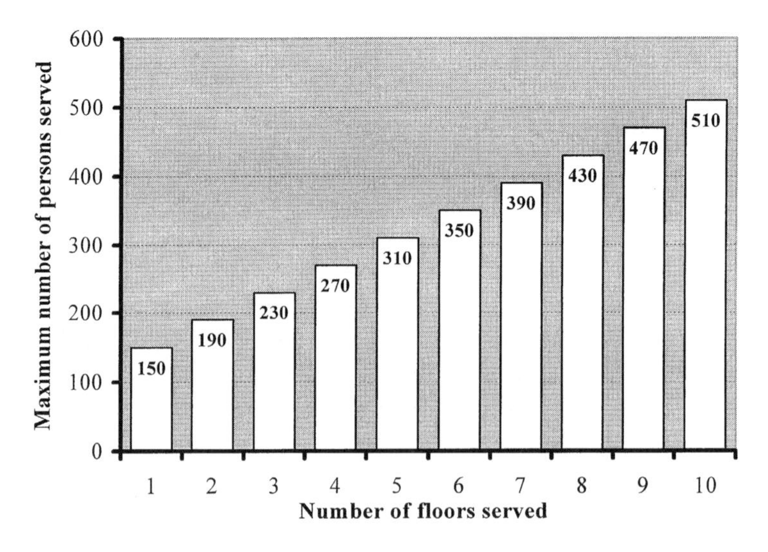

Fig. 5.5 Stair capacity for simultaneous evacuation – 1000 mm wide stairs.

In certain buildings it may be more advantageous to design stairs on the basis of 'phased evacuation'. Indeed, in high buildings it may be impractical or unnecessary to evacuate the building totally, especially if the recommendations regarding fire resistance, compartmentation and the installation of sprinklers are adhered to. In phased evacuation, people with reduced mobility and those most immediately affected by the fire (i.e. those people on the floor of fire origin and the one above it) are evacuated first. After that, if the need arises, floors can be evacuated two at a time. Phased evacuation allows narrower stairs to be used and has the added advantage that it causes less disruption in large buildings than total evacuation. Phased evacuation may be used for any buildings unless they are of the types listed above as needing total evacuation.

Where a building is designed for phased evacuation the following conditions should be met:

- The stairs should be approached through a protected lobby or protected corridor at each floor level. (This does not apply to a top storey.)
- The lifts should be approached through a protected lobby at each floor level.
- Each floor should be a compartment floor.
- If the building has a floor which is more than 30 m above ground, it should be protected throughout by an automatic sprinkler system which complies with the relevant requirements of BS 5306 : Part 2[7] (i.e. the sections dealing with the relevant occupancy rating and the additional requirements for life safety). This provision does not apply to flats of PG 1(a).
- An appropriate fire warning system should be fitted which complies with BS 5839 : Part 1[8] to at least the L3 standard.
- An internal speech communication system (such as a telephone, intercom system or similar) should be provided so that conversation is possible between a fire warden at every floor level and a control point at the fire service access level.
- Where it is deemed appropriate to install a voice alarm, the recommendations regarding phased evacuation in BS 5839 : Part 1 should be followed and the voice alarm system itself should conform to BS 5839 : Part 8[9].

When phased evacuation is used as the basis for design, the minimum stair width needed may be taken from Table 8 of AD B1 assuming phased evacuation of not more than two floors at a time. The data from Table 8, which has been reconfigured in Fig 5.6, is derived from the formula:

$$w = [(P \times 10) - 100] \text{ mm} \qquad \text{(formula 5.2)}$$

where
w = the minimum width of stair (w must not be less than 1000 mm), and
P = the number of people on the most heavily occupied storey.

5.2.8 *Step 8 – Escape stair design: protection*

To be effective as an area of relative safety during a fire, escape stairs need to have an adequate standard of fire protection. This relates not only to the

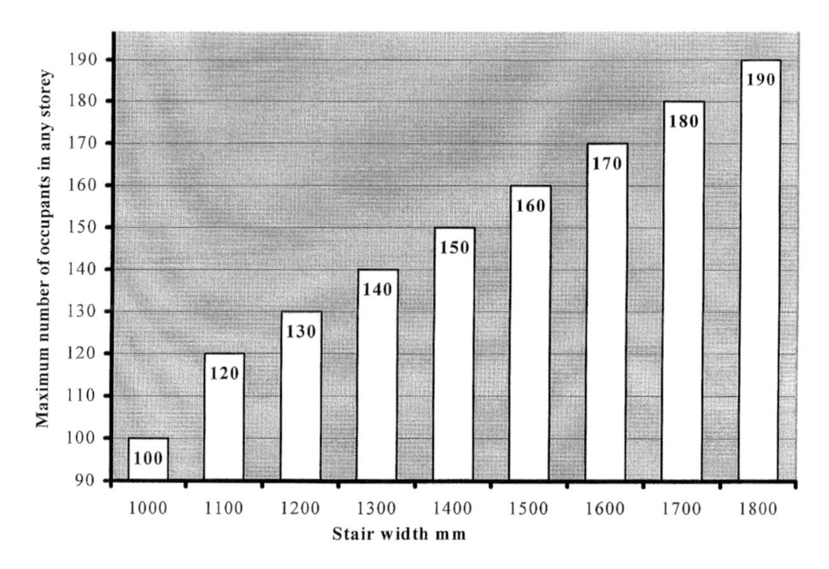

Fig. 5.6 Minimum width of stairs for phased evacuation.

presence of fire-resisting enclosures but also to the provision of protected lobbies, corridors and final exits. Therefore, each internal escape stair should be a protected stair situated in a fire-resisting enclosure. Additional measures may also be necessary for a stairway which is also a protected shaft (penetrating one or more compartment floors, see AD B section 9) or a fire-fighting shaft (see AD B section 18).

However, this does not preclude the provision of an accommodation stair (i.e. a stair which is provided for the convenience of occupants and is additional to those required for means of escape) if the design of the building so warrants it.

Exceptionally, an unprotected stair can form part of the internal escape route to a storey or final exit in low risk buildings if the number of people and the travel distance involved are very limited. For example, BS 5588 : Part 11 contains details in clause 10 of the use of an unprotected stair for means of escape in what are termed 'small premises'. Typically, the premises will be used as a small shop and will have two storeys consisting of:

- ground and first floor; *or*
- basement and ground floor.

The following conditions will also need to be complied with:

- the maximum floor area in any storey must not exceed 90 m^2;
- the maximum direct travel distance from any point in the ground storey to the final exit must not exceed 18 m;

- the maximum direct travel distance from any point in the basement or first storey to the stair must not exceed 12 m;
- the stair must deliver into the ground storey not more than 3 m from the final exit.

It should be noted that restaurant or bar premises and those used for the sale, storage or use of highly flammable materials are barred from this arrangement.

Generally, protected lobbies or corridors should be provided at all levels including basements (but not at the top storey) where:

- the building has a single stair and there is more than one floor above or below the ground storey (except for small premises – see above);
- the building has a floor which is more than 18 m above ground;
- the building is designed for phased evacuation;
- the option has been taken to not discount one stairway when calculating stair widths (see section 5.2.6).

In these cases an alternative to a protected lobby or corridor is the use of a smoke control system designed in accordance with BS 5588 : Part 4 as described earlier in section 5.2.5.

Protected lobbies are also needed where:

- the stairway is a fire-fighting stair (see AD B5);
- the stairway serves a place of special fire hazard (i.e. oil-filled transformer and switch gear rooms, boiler rooms, storage space for fuel or other highly flammable substances, rooms housing a fixed internal combustion engine, and – in schools – laboratories, technology rooms with open heat sources, kitchens and stores for PE mats or chemicals); in this case, the lobby should be ventilated by permanent vents with an area of at least 0.4 m^2, or should be protected by a mechanical smoke control system.

Since a protected stairway is considered to be a place of relative safety, it should be free of potential sources of fire. Therefore, the facilities that may be included in protected stairways are restricted to:

- washrooms or sanitary accommodation provided that the accommodation is not used as a cloakroom; the only gas appliances that may be installed are water heaters or sanitary towel incinerators;
- a lift well, on condition that the stairway is not a fire-fighting stair;
- an enquiry office or reception desk area of not more than 10 m^2, provided that there is more than one stair serving the building;
- fire-protected cupboards, provided that there is more than one stair serving the building;

- gas service pipes and meters, but only if the gas installation is in accordance with the requirements for installation and connection set out in the Pipelines Safety Regulations 1996[10] and the Gas Safety (Installation and Use) Regulations 1998[11].

If a protected stairway is situated on the external wall of a building it is not necessary for the external part of the enclosure to be fire-protected and in many cases it may be fully glazed. This is because fires are unlikely to start in protected stairways. Therefore these areas will not contribute to the radiant heat from a building fire which might put at risk another building.

In some building designs the stairway may be situated at an internal angle in the building façade (see Fig. 5.7) and may be jeopardised by smoke and flames coming from windows in the facing walls. This may also be the case if the stair projects from the face of the building (see Fig. 5.8) or is recessed into it. In these cases any windows or other unprotected areas in the face of the building and in the stairway should be separated by at least 1800 mm of fire-resisting construction. This provision also applies to flats and maisonettes.

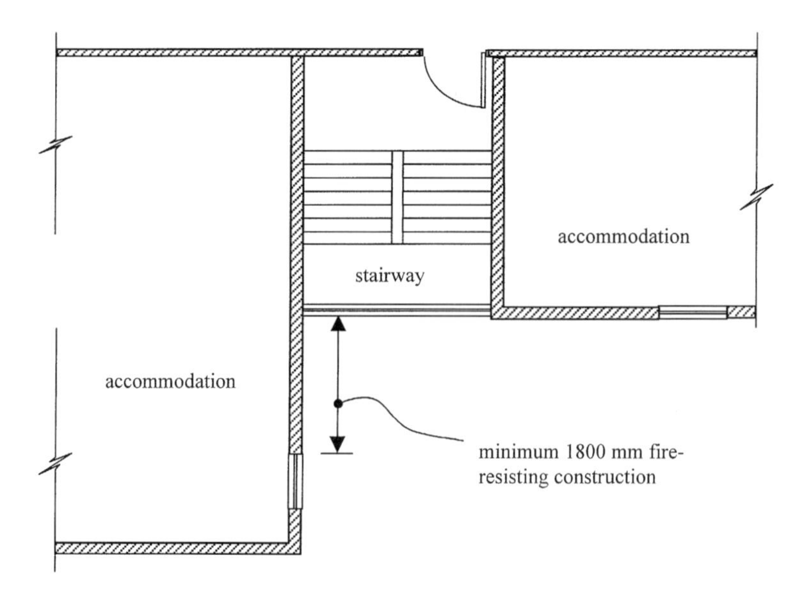

Fig. 5.7 External protection to protected stairway on internal angle.

5.2.9 Step 9 – Final exits

Ideally, every protected stairway should discharge directly to a final exit, i.e. it should be possible to leave the staircase enclosure and immediately reach a place of safety outside the building and away from the effects of fire.

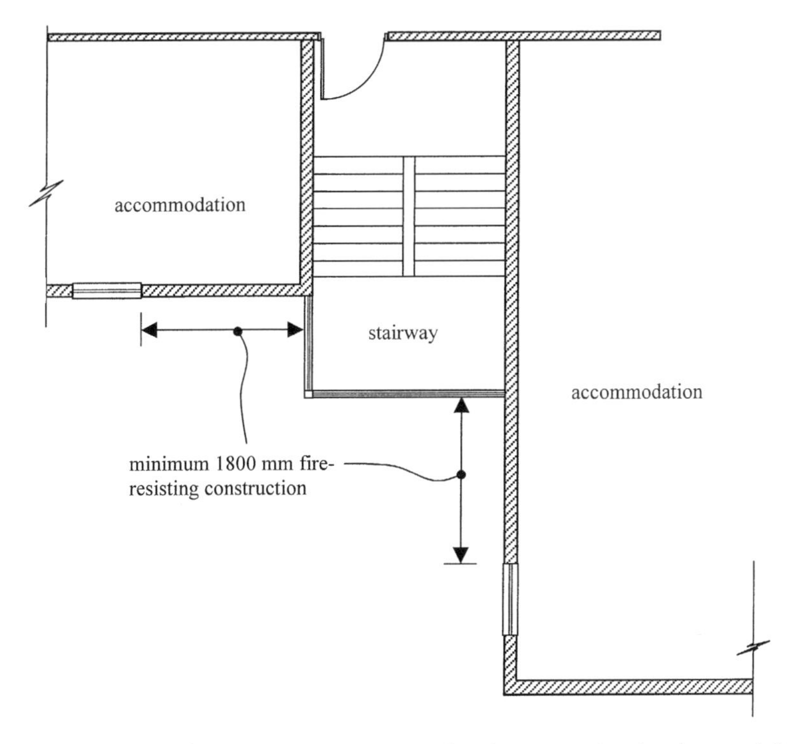

Fig. 5.8 External protection – protected stairway projecting beyond face of building.

Obviously, it is not always possible to achieve this ideal, especially where the building design calls for stairs to be remote from external walls. Therefore it is permissible for a protected stairway to discharge into a protected exit passageway, which in turn leads to a final exit from the building. Such a passageway can contain doors (e.g. to allow people on the ground floor to use the escape route) but they will need to be fire doors in order to maintain fire integrity. Therefore, if the exit route from a stair also picks up occupants of the ground and/or basement storeys, it may need to be increased in width accordingly. Thus, the width of the protected exit passageway will need to follow the guidance in step 4 above and be designed in accordance with Table 5.3, for the estimated numbers of people that will use it in an emergency.

Where a fire-fighting shaft is provided and the route from the foot of the shaft to the final exit is by way of a protected corridor, BS 5588 : Part 5[12] makes the following recommendations:

- the protected corridor, which should be considered to be part of the fire-fighting shaft, should not exceed 18 m in length;

- any access to the protected corridor from the accommodation should be by way of protected lobbies;
- it should not be necessary for persons escaping down the stair to pass through the fire-fighting lobby at fire service access level;
- to allow room for fire service personnel to move towards the fire-fighting shaft, where the corridor forms part of the means of escape from the accommodation it should be 500 mm wider than that required for means of escape purposes;
- the fire-fighting lobby should have a minimum area of 5 m^2 clear of any escape routes so that it can act as a fire service mustering point.

Exceptionally, it is permissible for the ground or basement stories of a dwellinghouse to discharge into an enclosed space such as a courtyard or back garden, provided certain conditions are met (see Chapter 6).

Sometimes the design of the building will call for two protected stairways or protected exit passageways to be adjacent to each other. Where this happens they should be separated by an imperforate enclosure.

Final exits should not be narrower than the escape routes they serve and should be positioned to facilitate evacuation of people out of and away from the building. This means that they should be:

- positioned so that rapid dispersal of people is facilitated to a street, passageway, walkway or open space clear of the effects of fire and smoke; the route from the building should be well defined and guarded if necessary;
- clearly apparent to users; this is very important where stairs continue up or down past the final exit level in a building;
- sited so that they are clear of the effects of fire from risk areas in buildings such as basements (e.g. outlets for basement smoke vents), and openings to transformer chambers, refuse chambers, boiler rooms and other similar risk areas.

5.2.10 Step 10 – External escape routes, lighting and signage

External escape stairs have long been used to provide additional vertical means of escape where parts of the building would otherwise be contained in long dead-ends or would be too distant from internal stairways. It is uncommon for such stairways to be fully protected from the elements, therefore they are not normally used for everyday access and egress around the building. Thus it can be argued that external escape stairs are not subject to the requirements of Part K of Schedule 1 to the Building Regulations 2000 because they do not form part of a building.

In buildings other than dwellings (for flats and maisonettes see Chapter 6), an external escape stair may be used as an alternative means of escape provided that there is more than one escape route available from a storey, or part of a building, and the following conditions are met:

- There is at least one internal escape stair available from every part of each storey (plant areas excluded).
- The external escape stair is not intended for use by members of the public if installed in assembly and recreation buildings (PG 5).
- It serves only office or residential staff accommodation if installed in an institutional building (PG 2(a))
- If it is more than 6 m in vertical extent it is sufficiently protected from adverse weather. This does not necessarily mean that full enclosure will be necessary. The stair may be located so that protection may be obtained from the building itself. In deciding on the degree of protection it is necessary to consider the height of the stair, the familiarity of the occupants with the building, and the likelihood of the stair becoming impassable as a consequence of adverse weather conditions.
- Any part of the building (including windows and doors etc.) which is within 1.8 m of the escape route from the stair to a place of safety should be protected with fire-resisting construction. This does not apply if there is a choice of routes from the foot of the stair, thereby enabling the people escaping to avoid the effects of fire in the adjoining building. Additionally, any part of an external wall which is within 1.8 m of an external escape route (other than a stair) should be of fire-resisting construction up to a height of 1.1 m from the paving level of the route.
- All the doors which lead onto the stair should be fire-resisting and self-closing. This does not apply to the only exit door to the landing at the head of a stair which leads downward.
- Any part of the external envelope of the building which is within 1.8 m of (and 9 m vertically below) the flights and landings of the stair, should be of fire-resisting construction. This 1.8 m dimension may be reduced to 1.1 m above the top landing level provided that this is not the top of a stair up from basement level to ground.
- Any glazing which is contained within the fire-resisting areas mentioned above should also be fire-resisting in terms of maintaining its integrity in a fire, and be fixed shut. (For example, Georgian wired glass is adequate; it does not also have to meet the requirements for insulation.)

Where more than one escape route exists from a storey or part of a building, one of those routes may be across a flat roof if the following conditions are observed:

- the route may not serve an institutional building (PG 2a) or any part of a building intended for use by members of the public;
- the flat roof should be part of the same building from which escape is being made;
- the escape route over the flat roof should lead to a storey exit or external escape route;
- the roof and its structure forming the escape route should be fire-resisting;
- any opening within 3 m of the route should be fire-resisting;
- the route should be adequately defined and guarded by walls and/or barriers in accordance with Approved Document K[13]. (This relates to the provision of barriers at least 1100 mm high designed to prevent people falling from the escape route. The barriers should be capable of resisting at least the horizontal force given in BS 6399 : Part 1[14].)

All escape routes should have adequate artificial lighting. In certain cases, escape lighting which illuminates the route if the mains supply fails, should also be provided (Table 5.6).

The lighting to escape stairs will also need to be on a separate circuit from that which supplies any other part of the escape route. Standards for installation of escape lighting systems are given in BS 5266 : Part 1[15], or CP 1007[16].

Except in dwellinghouses, flats and maisonettes, emergency exit signs should be provided to every escape route. It is not necessary to sign exits which are in ordinary, daily use. The exit should be distinctively and conspicuously marked by a sign with letters of adequate size complying with the Health and Safety (Safety Signs and Signals) Regulations 1996[17]. In general, these regulations may be satisfied by signs containing symbols or pictograms which are in accordance with BS 5499 : Part 1 and Part 4[18]. In some buildings other legislation may require additional signs.

5.3 General construction provisions

So far this chapter has dealt with the principles of design for means of escape, and has presented an evacuation model which will enable a designer to establish if the chosen layout for the building is feasible and will satisfy the minimum requirements of Part B1 of the Building Regulations.

This section takes the design one stage further by giving general guidance on a number of construction-related features of the means of escape which apply to all buildings, except dwellinghouses, concerning:

- the standard of protection necessary for the elements enclosing the means of escape;

Table 5.6 Provision of escape lighting.

Purpose group	Description of building or part	Areas where escape lighting is required	Areas where escape lighting is *not* required
1(a) 2(a) 2(b)	Flat or maisonette Institutional Other residential	All common escape routes (including external routes)	Common escape routes in two-storey flats. Dwellinghouses in PG 1(b) and 1(c)
3 4 6 7(a)	Office Shop and commercial[a] Industrial Storage and other non-residential	(i) Underground or windowless accommodation (ii) Stairways in a central core or serving storey(s) over 18 m from ground level (iii) Internal corridors more than 30 m long (iv) Open-plan areas exceeding 60 m^2	
4 7(b)	Shop and commercial[b] Car parks which admit the public	All escape routes (including external routes)	Escape routes in shops[c] of three or less storeys (with no sales floor exceeding 280 m^2)
5	Assembly and recreation	All escape routes (including external routes), and accommodation	(i) Accommodation open on one side to view sport or entertainment during normal daylight hours (ii) Parts of school buildings with natural light and used only during normal school hours
All	All	(i) Windowless toilet accommodation with floor area not exceeding 8 m^2 (ii) All toilet accommodation with floor area exceeding 8 m^2 (iii) Electricity and generator rooms (iv) Switch room/battery room for emergency lighting system (v) Emergency control room	Dwellinghouses in PG 1(b) and 1(c)

[a] Those parts of the premises where the public are not admitted.
[b] Those parts of the premises where the public are admitted
[c] Any 'shop' (see definition in section 4.2) which is a restaurant or bar will require escape lighting as indicated in column 3.

- the provision of doors;
- the construction of escape stairs;
- mechanical services including lift installations;
- protected circuits for the operation of equipment in the event of fire;
- refuse chutes and storage;
- the provision of fire safety signs.

These recommendations should be read in conjunction with the provisions described in Chapter 7 for different building types (except dwellinghouses).

5.3.1 Protection of escape routes – standards and general constructional provisions

Those parts of a means of escape which are required by Part B1 to be fire-resisting should comply with the recommendations given in AD B3 (Internal fire spread – structure) or AD B5 (Access and facilities for the fire service) in addition to AD B1. In most cases 30 minutes' fire protection is sufficient for the protection of a means of escape. The exceptions to this are when the element also performs a fire-separating function, or separates areas of different fire risk, such as:

- a compartment floor;
- a compartment wall;
- an external wall;
- a protected shaft; *or*
- a fire-fighting shaft.

In these cases the element should achieve the standard of fire resistance given in Table A2 of Appendix A to Approved Document B. This may require considerably more than 30 minutes' fire resistance (the fire resistance periods in Table A2 range from 30 minutes to 120 minutes).

Details of the fire resistance recommendations for those parts of the building that have a direct bearing on the means of escape are given in Table 5.7. This is extracted from Table A1 of Appendix A to Approved Document B.

As may be seen from Table 5.7, glazed elements in fire-resisting enclosures and doors which are only able to meet the requirements for integrity in the event of a fire, will be limited in area to the amounts shown in Table A4 of Appendix A of AD B.

There are no limitations on the use of glazed elements that can meet both the integrity and insulation performance recommendations of AD B1. However, there may be some restrictions on the use of glass in fire-fighting stairs and lobbies in BS 5588 : Part 5 under the recommendations for robust construction.

Table 5.7 Fire resistance of elements forming the means of escape.

Element	Description	Minimum fire resistance period (minutes) when tested to the relevant part of BS 476			Method of exposure in BS 476 test
		Loadbearing capacity[a]	Integrity[b]	Insulation[c]	
1. Roof	Forming part of an escape route	30	30	30	From underside[d]
2. External wall	Any part adjacent to an external escape route	30	30	No provision[e]	From inside the building
3. Protected shaft	(a) Glazing[f] in a screen separating a corridor from a protected shaft	Not applicable	30	No provision	Each side separately
	(b) Unglazed part of the screen described above	30	30	30	
4. Enclosure	(a) To a protected stairway[g,h]	30	30	30	Each side separately
	(b) To a lift shaft[g]				
	(c) To a protected lobby[h]				
	(d) To a protected corridor[h]				
	(e) To a protected entrance hall or protected landing in a flat or maisonette[h]				
5. Sub-division	Of a corridor separating alternative escape routes[h]	30	30	30	Each side separately
6. Fire-resisting construction	Enclosing a place of special fire hazard[h]	30	30	30	Each side separately

Notes:

[a] The element must not suffer structural collapse within the given time period when tested to BS 476.
[b] The element must resist fire penetration within the given time period when tested to BS 476.
[c] The element must resist the transfer of excessive heat by conduction within the given time period when tested to BS 476.
[d] A suspended ceiling can contribute to the overall assessment of fire resistance if it complies with the appropriate provisions of Table A3 of Appendix A to Approved Document B.
[e] Realistically, this provision only applies to any unprotected areas (windows, doors, etc.) in the external wall, since in most cases the wall will need to achieve the fire resistance period recommended for the building as a whole.
[f] The area of any glazing in these elements may be limited. See Table A4 of Appendix A to Approved Document B.
[g] Only applies if the enclosure is not considered to be a protected shaft or compartment wall.
[h] See section 5.2.8 for definition.

The floors of escape routes, including the surfaces of steps and ramps, should be chosen so that they are not unduly slippery when wet. Additionally, sloping floors or tiers should not have a pitch greater than 35° to the horizontal.

Where a ramp forms part of an escape route it should also meet the provisions of Approved Document M[19]. This means that it should ideally be not steeper than 1 in 20. However, AD M allows a 1 in 15 ramp if the individual flight is no longer than 10 m and a 1 in 12 ramp is allowed where the flight length does not exceed 5 m.

From the aspect of safety in use, further guidance on the provision of ramps and associated landings, and on aisles and gangways in places where there is fixed seating, may be found in Approved Document K.

5.3.2 *The provision of doors on escape routes*

Doors on escape routes often need to be fire-resisting. This means that certain test criteria and performance standards as set out in Appendix B of Approved Document B will need to be met. Table 5.8 (which is based on Table B1 to Appendix B) shows where doors should be provided on means of escape and gives details of their fire resistance performance.

Table 5.8 Provision of fire doors on means of escape.

Location of door	Minimum fire resistance of door (minutes)[a]	Smoke seals[b] needed?
1. In a compartment wall:		
(a) separating buildings	As for the wall in which fitted but minimum FD60	
(b) separating a flat or maisonette from a space in common use	FD30	✓
(c) enclosing a protected shaft to a stairway wholly or partly above ground level in PG 1(a) Flats; PG 2(b) Other Residential; PG 3 Office; or PG5 Assembly and Recreation	FD30	✓
(d) enclosing a protected shaft to a stairway in all other Purpose Groups	The greater of half the fire resistance period needed for the wall in which located, or FD30	✓
(e) enclosing a protected shaft to a lift or service shaft	The greater of half of the fire resistance period needed for the wall in which located, or FD30	
(f) not described in 1(b) to (e) above	As for the wall in which fitted	
(g) As (f) and used for progressive horizontal evacuation	As for the wall in which fitted	✓

Cont.

Table 5.8 Continued.

Location of door	Minimum fire resistance of door (minutes)[a]	Smoke seals[b] needed?
2. Forming part of the enclosures of:		
(a) a protected stairway (but not one which forms a protected shaft or is in a single family dwelling)	FD30	✓
(b) a lift shaft (but not one which forms a protected shaft)	FD30	
(c) a protected lobby approach (or protected corridor) to a stairway	FD30	✓
(d) any other protected corridor	FD20	✓
(e) a protected lobby approach to a lift shaft	FD30	✓
(f) a protected entrance hall or protected landing in a flat or maisonette	FD20	
3. Giving access:		
(a) to an external escape route	FD30	
(b) within a cavity barrier	FD30	
(c) between a dwellinghouse and an attached garage	FD30	
4. Sub-dividing:		
(a) corridors which connect alternative exits	FD20	✓
(b) dead-end portions of corridors from the remainder of the corridor	FD20	✓
5. In any fire-resisting construction not described elsewhere in this table	FD20	

[a] The door must resist fire penetration within the given time period (i.e. it must maintain its integrity) when tested to BS 476 : Part 22[21]. For example, a door which satisfies the test criteria for 30 minutes is termed an FD30 door. Since doors offer very little in terms of insulation, the proportion of door openings in compartment walls is limited to 25% of the length of the compartment wall unless the doors can provide adequate resistance to both integrity and insulation.

[b] The doors marked with a tick (✓) in this column are sited where it is necessary to resist smoke leakage at ambient temperatures. The leakage rate should not exceed 3 m^3/m/hour (head and jambs only) when tested at 25 Pascals. Such doors are identified using the suffix S when specified as part of a design (see case study in Appendix A of this book).

Further advice on the specification, design, construction, installation and maintenance of fire doors constructed with non-metallic door leaves can be obtained from BS 8214[20].

The time taken to pass through a closed door can be critical when escaping from a building in a fire situation. Doors on escape routes should be readily openable if undue delay in escaping from a building is to be avoided. A number of factors, which are relevant to the use of doors, are set out below.

In general, escape doors should open in the direction of the means of escape where it is reasonably practicable to do so and should always do so if more than 60 people are likely to use the door in an emergency. However, for some industrial activities where there is a very high risk with potential for rapid fire growth, it may be necessary for escape doors to open in the direction of escape for lower occupant numbers than 60. The exact figure will depend on the individual circumstances of the case and there is no specific guidance laid down in AD B.

Ideally, doors on escape routes (whether or not they are fire doors) should not be fitted with fastenings unless these are simple to use and can be operated from the side of the door which is approached by people escaping. Any fastenings should be able to be operated without a key and without having to operate more than one mechanism; however this does not prevent doors being fitted with ironmongery which allows them to be locked when the rooms are empty. For example, this would permit a hotel bedroom to be fitted with a lock which could be operated from the outside with a key and from the inside by a knob or lever.

No fire door should be hung on hinges made of a material with a melting point less than 800°C, unless the hinges can be shown to be satisfactory when tested as part of the door assembly.

Where security of final exit doors is important, as in Assembly and Recreation (PG 5) and Shop and Commercial (PG 4) buildings, panic bolts may be used. Additionally, it is accepted that in non-residential buildings it is appropriate for final exit doors to be locked when the building is empty. Clearly, a good deal of responsibility must be placed on management procedures for the safe use of these locks.

Ironmongery used on fire doors can significantly affect their performance in a fire. Reference should be made to the *Code of Practice, Hardware essential to the optimum performance of fire-resisting timber doorsets*, published by the Association of Builders' Hardware Manufacturers, 1993, where guidance may be obtained on the selection of suitable hardware.

Recommendations for self-closers and hold-open devices for fire doors are contained in Appendix B of AD B. These recommendations may be summarised as:

(1) All fire doors should be fitted with an automatic self-closing device unless they are to cupboards or service ducts which are normally kept locked shut.
(2) Rising butt hinges are not considered to be automatic self-closing devices but may be used on a door to (or within) a dwelling, in a cavity barrier, or between a dwellinghouse and a garage.
(3) In some cases (such as in hospital corridors) a self-closing device may be considered to be a hindrance to normal use. In such cases self-closing fire doors may be held open by:
    - A fusible link. (Generally, this option does not apply to doors which are fitted in an opening on a means of escape. The one exception to

this is the case where two self-closing fire doors are fitted in the same opening on a means of escape such that the required level of fire resistance can be achieved by the two doors together. In such a case, the opening must be capable of being closed by either door and one of them may be fitted with a fusible link. The other door must be capable of being easily opened by hand and have at least 30 minutes' fire resistance.)

– A door closure delay device.
– An automatic release mechanism actuated by an automatic fire detection and alarm system. (In this context an automatic release mechanism is a device which automatically closes a door in the event of each or any one of:
  - smoke detection by suitable automatic apparatus;
  - manual operation by a suitably located switch;
  - failure of the electricity supply to the device, smoke detection apparatus or manual switch;
  - operation of a fire alarm system, where fitted.)

Sometimes it is necessary to install fire-resisting roller shutters across an opening in a compartment wall or floor, which will close automatically, in order to maintain the integrity of compartmentation in a fire situation. Clearly, if such shutters were installed across a means of escape, it would be possible for occupants to become trapped if the shutter closed before that part of the building had been evacuated. To avoid this, shutters across a means of escape should only be released by a heat sensor (e.g. a fusible link or electric heat detector) in the immediate vicinity of the door. Smoke detectors or fire alarm systems should not be used to initiate closure of fire shutters in such situations unless the shutter is also designed to partially descend to form part of a boundary to a smoke reservoir.

Doors on escape routes should swing through at least 90° to open, and should not reduce the effective width of any escape route across a landing. The swing should be clear of any changes in floor level, although a single step or threshold on the line of a door opening is permitted. Any door that opens towards a corridor or stairway should be recessed so that it does not encroach on or reduce the effective width of the corridor or stairway.

Doors on escape routes which sub-divide corridors, or are hung to swing in two directions, should contain vision panels. (There are also provisions in Approved Document M for vision panels in doors across accessible corridors.)

If revolving or automatic doors, or turnstiles, are placed across an escape route it is possible that they might obstruct the passage of people escaping. Therefore, they should not be placed across an escape route unless:

(1) in the case of automatic doors which are the correct width for the design of the route they:
   - will fail safely to become outward opening from any position of opening; *or*
   - are provided with a monitored failsafe system for opening the doors in the event of mains power failure; *or*
   - fail safely in the open position in the event of mains power failure; *or*

(2) they have non-automatic swing doors of the required width adjacent to them which can provide an alternative exit.

5.3.3 The construction of escape stairs – conventional stairs

Escape stairs and their associated landings in certain high risk situations or buildings require the extra safeguard of being constructed in materials of limited combustibility. These are composite materials (such as plasterboard) which include combustible materials in their composition so that they cannot be classed as totally non-combustible. When exposed as linings to walls or ceilings they must achieve certain low flame spread ratings.

This recommendation applies in the following cases:

- where a building has only one stair serving it (this does not apply to two and three-storey flats and maisonettes);
- where a stair is located in a basement storey (except if it is a private stair in a maisonette);
- to any stair serving a storey in a building which is more than 18 m above ground or access level;
- to any external stair (except where it connects the ground floor or paving level to a floor or flat roof which is not more than 6 m above ground);
- if the stair is a fire-fighting stair.

In all the above, except for the fire-fighting stair, it is permissible to add combustible materials to the upper surface of the stair.

Where possible, single steps should be avoided on escape routes unless prominently marked, since they can cause falls. It is permissible though, to have a single step on the line of a doorway.

5.3.4 The construction of escape stairs – special stairs and ladders

Although spiral and helical stairs and fixed ladders are not as inherently safe as conventional stairs they may be used as part of a means of escape if the following restrictions are observed:

- Spiral and helical stairs should be designed in accordance with BS 5395 : Part 2[22]. They are not suitable for use by pupils in schools, and if used by members of the public they should be type E (public) stair from the above standard.
- Fixed ladders are not suitable as a means of escape for members of the public. They should only be used to access areas which are not normally occupied, such as plant rooms, where it is not practical to provide a conventional stair. They should be constructed of non-combustible materials.

5.3.5 Mechanical services including lift installations

Mechanical ventilation systems should be designed so that in a fire:

- air is drawn away from protected escape routes and exits; *or*
- the system (or the appropriate part of it) is closed down.

Systems which recirculate air should comply with BS 5588 : Part 9[23], for operation under fire conditions. Recommendations for the use of mechanical ventilation in a place of assembly are given in BS 5588 : Part 6[24]. Guidance on the design and installation of mechanical ventilation and air-conditioning plant is given in BS 5720[25].

Where a pressure differential system is installed in a building (in order to keep smoke from entering the means of escape) it should be compatible with any ventilation or air-conditioning systems in the building, when operating under fire conditions.

Lifts are not normally used for means of escape since there is always the danger that they may become immobilised due to power failure and may trap the occupants. It is possible to provide lifts as part of a management plan for evacuating disabled people if the lift installation is appropriately sited and protected. It should also contain sufficient safety devices to ensure that it remains usable during a fire. Further details may be found in BS 5588 : Part 8[26]. This is considered more fully in Chapter 7.

A further problem with lifts is that they connect floors and may act as a vertical conduit for smoke or flames, thus prejudicing escape routes. This may be prevented if the following recommendations are observed:

- Lift wells should be enclosed throughout their height with fire-resisting construction if their siting would prejudice an escape route. Alternatively, they should be contained within the enclosure of a protected stairway.
- Any lift well which connects different compartments in a building should be constructed as a protected shaft.

- In buildings where escape is based on the principles of phased or progressive horizontal evacuation, if the lift well is not within the enclosure of a protected stairway, its entrance should be separated from the floor area on each storey by a protected lobby.
- Similarly, unless the lift is in a protected stairway enclosure, it should be approached through a protected lobby or corridor:
  - if it is situated in a basement or enclosed car park; or
  - where the lift delivers directly into corridors serving sleeping accommodation if any of the storeys also contain high fire risk areas such as kitchens, lounges or stores.
- A lift should not continue down to serve a basement if there is only one escape stairway in the building (since smoke from a basement fire might be able to prejudice the escape routes in the upper storeys) or if it is in an enclosure to a stairway which terminates at ground level.
- Lift machine rooms should be located over the lift shaft wherever possible. Where a lift is within the only protected stairway serving a building and the machine room cannot be located over the lift shaft, then it should be sited outside the protected stairway. This is to prevent smoke from a fire in the machine room from blocking the stair.
- Wall-climber and feature lifts often figure in large volume spaces such as open malls and atria. Such lifts do not have a conventional well and may place their occupants at risk if they pass through a smoke reservoir. Care will be needed in the design in order to maintain the integrity of the reservoir and protect the occupants of the lift.

5.3.6 Protected circuits for the operation of equipment in the event of fire

Protected power circuits are provided in situations where it is critical that the circuit should continue to function during a fire. For example, this will apply where the circuits provide power to:

- fire extinguishing systems;
- smoke control systems;
- sprinkler systems;
- fire-fighting shaft systems (such as fire-fighting lifts);
- motorised fire shutters;
- CCTV systems installed for monitoring means of escape;
- data communications systems that link fire safety systems.

The cable used in a protected power circuit for operation of equipment in the event of fire should:

- meet the requirements for classification as CWZ in accordance with BS 6387[27];
- follow a route which passes through parts of the building in which there is negligible fire risk;
- be separate from circuits which are provided for other purposes.

5.3.7 *Refuse chutes and storage*

Fires in refuse chute installations are extremely common and these installations are required in Approved Document B3 to be built of non-combustible materials. So that escape routes are not jeopardised, refuse chutes and rooms for refuse storage should:

- be separated, by fire-resisting construction, from the rest of the building; *and*
- not be located in protected lobbies or stairways.

Rooms which store refuse or contain refuse chutes should:

- be approached directly from the open air; *or*
- be approached via a protected lobby provided with at least $0.2\,m^2$ of permanent ventilation.

Refuse storage chamber access points should be sited away from escape routes, final exits and windows to dwellings.

Refuse storage chambers, chutes and hoppers should be sited and constructed in accordance with BS 5906[28].

5.3.8 *The provision of fire safety signs*

Doors which are to be kept closed or locked when not in use, or which are held open by an automatic release mechanism, should be marked with the appropriate fire safety sign in accordance with BS 5499. The signs should be on both sides of the fire doors, except for cupboards and service ducts where it is only necessary to mark the doors on the outside. This recommendation does not apply to:

- fire doors within dwellinghouses;
- fire doors to and within flats or maisonettes;
- bedroom doors in Residential (other) buildings (PG 2(b));
- lift entrance doors.

5.4 References

1 BS 5588: Fire precautions in the design, construction and use of buildings: Part 10: 1991 Code of practice for shopping complexes.
2 Firecode. HTM 81. Fire precautions in new hospitals. (NHS Estates) HMSO, 1996 and HTM 88. Guide to fire precautions in NHS housing in the community for mentally handicapped (or mentally ill) people. (DHSS) HMSO, 1986.
3 *Design methodologies for smoke and heat exhaust ventilation.* BR 368, BRE 1999 Revision of *Design principles for smoke ventilation in enclosed shopping centres.* BR 186, BRE 1990.
4 *Fire Precautions Act 1971. Guide to fire precautions in existing places of work that require a fire certificate. Factories, offices, shops and railway premises.* (Home Office/Scottish Office) HMSO, 1993.
5 BS 5588: Fire precautions in the design, construction and use of buildings: Part 11: 1997 Code of practice for shops, offices, industrial, storage and other similar buildings.
6 BS 5588: Fire precautions in the design, construction and use of buildings: Part 4: 1998 Code of practice for smoke control using pressure differentials.
7 BS 5306 Fire extinguishing installations and equipment on premises, Part 2: 1990 Specification for sprinkler systems.
8 BS 5839 Fire detection and alarm systems for buildings, Part 1: 1988 Code of practice for system design, installation and servicing.
9 BS 5839: Fire detection and alarm systems for buildings, Part 8: 1998 Code of practice for the design, installation and servicing of voice alarm systems.
10 Pipelines Safety Regulations 1996, SI 1996/825.
11 Gas Safety (Installation and Use) Regulations 1998 SI 1998/2451.
12 BS 5588: Fire precautions in the design, construction and use of buildings Part 5 : 1991 Code of practice for fire-fighting stairways and lifts.
13 Approved Document K, Protection from falling, collision and impact. DTLR and the Welsh Office 1998.
14 BS 6399 Loadings for buildings, Part 1: 1984 Code of practice for dead and imposed loads.
15 BS 5266 Emergency lighting, Part 1: 1988 Code of practice for the emergency lighting of premises other than cinemas and certain other specified premises used for entertainment.
16 CP 1007: 1955 Maintained lighting for cinemas.
17 Health and Safety (Safety Signs and Signals) Regulations 1996 (SI 1996/2341).
18 BS 5499 Fire safety signs, notices and graphic symbols, Part 1: 1990 Specification for fire safety signs. Part 4: 2000 Code of practice for escape route signing.
19 Approved Document M Access and facilities for disabled people. DETR and the Welsh Office 1999.
20 BS 8214: 1990 Code of practice for fire door assemblies with non-metallic leaves.
21 BS 476: Fire tests on building materials and structures Part 22: 1987 Methods for determination of fire resistance of non-loadbearing elements of construction.

22 BS 5395 Stairs, ladders and walkways, Part 2: 1984 Code of practice for the design of helical and spiral stairs.
23 BS 5588: Fire precautions in the design, construction and use of buildings Part 9: 1989 Code of practice for ventilation and air conditioning ductwork.
24 BS 5588: Fire precautions in the design, construction and use of buildings Part 6: 1991 Code of practice for assembly buildings.
25 BS 5720: 1979 Code of practice for mechanical ventilation and air conditioning in buildings.
26 BS 5588: Fire precautions in the design, construction and use of buildings Part 8: 1988 Code of practice for means of escape for disabled people.
27 BS 6387: 1994 Specification for performance requirements for cables required to maintain circuit integrity under fire conditions.
28 BS 5906: 1980 Code of practice for storage and on-site treatment of solid waste from buildings.

Chapter 6
Dwellinghouses, flats and maisonettes

6.1 Introduction

Chapter 4 contains details of the general principles for the provision of means of warning and escape which apply to all building types. Additionally, in Chapter 5, a simple evacuation model is given for the design of means of escape in other residential buildings (hotels, hostels, halls of residence, small nursing homes and other similar institutional buildings where people sleep on the premises) and smaller, simpler buildings used for other non-residential purposes. This is based on the recommendations of sections 1 to 6 of Approved Document B.

Because of its generality, the evacuation model outlined in chapter 5 may prove to be too detailed for the design of means of escape in dwellings, whilst at the same time proving insufficient for the needs of more complex buildings. The purpose of this chapter is to consider means of escape issues that are specific to dwellinghouses, flats and maisonettes and to supplement the guidance in Approved Document B1 (AD B1) with sources of information which are referred to, but are not actually contained within that document.

Dwellinghouses

For the purposes of Building Regulations a dwellinghouse means a dwelling which is not a flat or maisonette. At its most basic level it will be a house or bungalow occupied by a single family (or a group of people living as a single household). Over the years, a good deal of argument has taken place in the law courts regarding the parameters which determine what constitutes a 'single household'. For example, a group of students occupying a house, which has been let to them by a university and which has to be vacated when they leave the university, may be regarded as a single household (i.e. they are not living in a house in multiple occupation, see Chapter 7).

Clearly, there is a greater degree of risk from fire to the occupants where they are unrelated strangers than if they are part of a single household, since more control may be exercised over the use of the dwelling in the latter case. Therefore, it is most important to consider the different modes of use of dwellinghouses when designing suitable means of warning and escape.

6.2 Fire alarm and detection systems in dwellinghouses

In most dwellinghouses, safety levels can be significantly increased by installing systems which automatically give early warning of fire. Approved Document B1 gives the following range of options but fails to indicate the circumstances in which they might be used:

- an automatic fire detection and alarm system in accordance with the relevant recommendations of BS 5839 Part 1[1]: 1988 to at least the L3 standard specified in the code; *or*
- an automatic fire detection and alarm system in accordance with the relevant recommendations of BS 5839 Part 6[2]: 1995 to at least a Grade E type LD3 standard; *or*
- a suitable number of smoke alarms provided in accordance with the guidance given in AD B1.

BS 5839 : Part 6 applies to dwellinghouses, flats and maisonettes, mobile homes, sheltered housing, NHS housing in the community for mentally handicapped or mentally ill people (see also HTM 88[3]) and houses divided into several self-contained single-family dwelling units. It does not apply to hostels, caravans or boats (except permanently moored boats used solely as living accommodation), or to the communal parts in purpose-built sheltered housing and blocks of flats or maisonettes. Table 2 of BS 5839 : Part 6 gives a wider range of options for the minimum grade and type of fire detection and alarm system for protection of life in typical dwellings. Table 6.1 in this chapter is derived from Table 2 and should be read in conjunction with the definitions which follow here.

Grade

A – fire detection and alarm system incorporating control and indicating equipment which conforms to BS 5839 : Part 4[6], designed, installed and serviced in accordance with most of the recommendations of BS : 5839 Part 1.
B – fire detection and alarm system comprising fire detectors other than smoke

Table 6.1 Minimum grade and type of fire detection and alarm system for protection of life in typical dwellings.

Class of dwelling	New dwellings conforming to the recommendations of BS 5588 : Part 1	
	Grade	Type
Single-family dwellings with no floor greater than 200 m² in area		
Bungalow, flat or other single-storey unit	E	LD3
Owner-occupied maisonette or house of two storeys, with no floor level exceeding 4.5 m in height above ground	E	LD3
Rented maisonette or house of two storeys, with no floor level exceeding 4.5 m in height above ground	E	LD3
House in which one floor level is more than 4.5 m above ground	C	LD3
House in which more than one floor level is more than 4.5 m above ground.	B	LD2[a]
Single-family dwellings with one or more floors greater than 200 m² in area		
Single-storey dwelling	C	LD3
Two-storey house with no floor level exceeding 4.5 m in height above ground	B	LD3
House in which one or more floor levels are more than 4.5 m above ground	Grade A, with detectors sited in accordance with the BS 5839 : Part 1 for a type L2[ab] system	
Houses in multiple occupation		
Dwelling of one or two storeys with no floor level exceeding 4.5 m in height above ground and no floor greater than 200 m² in area	Not applicable[c]	Not applicable[c]
Other dwellings:		
Individual dwelling units comprising two or more rooms	Not applicable[c]	Not applicable[c]
Communal areas	Not applicable[c]	Not applicable[c]
Sheltered housing (individual dwelling units only)[d]	C	LD3
NHS housing in the community for mentally handicapped or mentally ill people		
Dwellings of one or two storeys occupied by no more than seven mentally handicapped or mentally ill residents	B	LD1[e]
Other dwellings	Grade A, with detectors sited in accordance with the recommendations of BS 5839 : Part 1 for a type L1 system	

[a] Detectors should be installed in any communal kitchen and in any communal living room.
[b] BS 5839 : Part 1 recommends that detectors are installed in escape routes and rooms adjoining escape routes. Notwithstanding the recommendations of BS 5839 : Part 1, detectors may be omitted from rooms adjoining escape corridors of 6 m or less in length.
[c] New houses in multiple occupation can be treated as normal dwellings if the number of occupants does not exceed six. Above this figure the guidance given in DoE Circular 12/94[4] or Welsh Office Circular 25/92[5] may be used.
[d] In such dwellings individual dwelling units will be connected to a central monitoring point so that the person in charge is aware that a fire has been detected in one of the dwellings and can identify the dwelling concerned.
[e] Type LD2 may be acceptable for a single-storey building which has an alternative means of escape to a separate exit from the dwelling.

alarms, fire alarm sounders, and control and indicating equipment which either conforms to BS 5839 : Part 4 or to Annex B of BS 5839 : Part 6.
C – a system of fire detectors and alarm sounders (which may be combined in the form of smoke alarms) connected to a common power supply, comprising the normal mains and a standby supply, with an element of central control.
D – a system of one or more mains-powered smoke alarms, each with an integral standby supply.
E – a system of one or more mains-powered smoke alarms with no standby supply.
F – a system of one or more battery-powered smoke alarms.

Type

The designation 'LD' is used to distinguish automatic fire detection and alarm systems intended for the protection of life in dwellings, from type L systems as defined in BS : 5839 Part 1, which are intended for the protection of life in any type of building.

LD1 – a system installed throughout the dwelling, which incorporates detectors in all circulation spaces that form part of the escape routes from the dwelling. The detectors are also placed in all rooms and areas in which fire might start, other than toilets, bathrooms and shower rooms.
LD2 – a system incorporating detectors in all circulation spaces that form part of the escape routes from the dwelling, and in all rooms or areas that present a high fire risk to occupants (normally living rooms, dining rooms and bedrooms, but see clause 4 of BS 5839 : Part 6).
LD3 – a system incorporating detectors in all circulation spaces that form part of the escape routes from the dwelling.

The guidance given in AD B1 is summarised below since it is based on the principal recommendations of the codes referred to above. For most dwellinghouses in normal use it represents a suitable minimum standard.

6.2.1 Smoke alarms

Smoke alarms should be:

- Mains-operated, and designed and manufactured in accordance with BS 5446 : Part 1[7]: 1990. They may have a secondary power supply (such as a capacitor or rechargeable or replaceable battery – see clause 13 of BS 5839 : Part 6).

- Located in circulation areas between sleeping places and places where fires are likely to start (kitchens and living rooms) and within 7.5 m of the door to every habitable room.
- Fixed to the ceiling and at least 300 mm from any walls and light fittings (unless there is test evidence to prove that the detector will not be adversely affected by the proximity of a light fitting). Units specially designed for wall mounting are acceptable provided that they are mounted above the level of doorways into the space and are fixed in accordance with manufacturers' instructions.
- Sited so that the sensor, for ceiling mounted devices, is between 25 mm and 600 mm below the ceiling (between 25 mm and 150 mm for heat detectors), assuming that the ceiling is predominantly flat and level.
- Fixed in positions that allow for routine maintenance, testing and cleaning (i.e. not over a stairwell or other floor opening).
- Sited away from areas where steam, condensation or fumes could give false alarms (this would include heaters, air-conditioning outlets, bathrooms, showers, cooking areas or garages, etc.).
- Sited away from areas that get very hot (e.g. boiler rooms) or very cold (e.g. an unheated porch). They should not be fitted to surfaces which are either much hotter or much colder than the rest of the room since air currents might be created which would carry smoke away from the unit.

The number of alarms which are installed will depend on the size and complexity of the layout of the dwelling and should be based on an analysis of the risk to life from fire. The following minimum provisions should be observed:

- At least one alarm should be installed in each storey of the dwelling.
- Where more than one smoke alarm is installed they should be interconnected[8] so that the detection of smoke in any unit will activate the alarm signal in all of them. Manufacturers' instructions should be adhered to regarding the maximum number of units that can be interconnected.
- In open plan designs where the kitchen is not separated from the stairway or circulation space by a door, the kitchen should contain a compatible interlinked heat detector. This should be in addition to the normal provision of smoke detector(s) in the circulation space(s).
- Where there are long hallways or corridors, additional smoke detectors may be necessary.

It should be noted that maintenance of the system in use is of utmost importance. Since this cannot be made a condition of the passing of plans for Building Regulation purposes it is important to ensure that developers and builders provide occupiers with full details of the use of the equipment and its maintenance (or guidance on suitable maintenance contractors). BS 5839 : Parts 1 and 6 also recommend that occupiers receive manufacturers' operating and maintenance instructions.

Power supplies to smoke alarms[9]

Power supplies for smoke alarm systems should be derived from the dwelling's mains electricity supply and connected to the smoke alarms through a separately fused circuit at the distribution board (consumer unit).

The smoke alarm system can include a stand-by power supply which will operate during mains failure. This can allow the system to obtain its power by connection to a regularly used local lighting circuit with the advantage that the circuit will be unlikely to be disconnected for any prolonged period. Where the system does not include a stand-by power supply, no other electrical equipment should be connected to the smoke alarm circuit (except for a mains failure monitoring device).

The mains supply to the smoke alarm system can be monitored by a device which will give warning in the event of failure of the supply. The warning of failure may be visible or audible (in which case it should be possible to silence it) and should be sited so that it is readily apparent to occupants. The circuit for the mains failure monitor should be designed so that any significant reduction in the reliability of the mains supply is avoided.

Ideally, the smoke alarm circuit should not be protected by any residual current device (rcd) such as a miniature circuit breaker or earth leakage trip. However, sometimes it is necessary for reasons of electrical safety that such devices be used and in these cases, either:

- the rcd should serve only the circuit supplying the smoke alarms; *or*
- the rcd protection of a fire alarm circuit should operate independently of any rcd protection for circuits supplying socket outlets or portable equipment.

Since it does not need any fire survival properties, the mains supply to smoke alarms, and any interconnecting wiring, may comprise any cable which is suitable for ordinary domestic mains wiring. Cables used for interconnections should be readily identifiable from those supplying power (e.g. by colour coding). Cables should be protected against damage in any areas where they may be subject to impact, abrasion or rodent attack.

6.2.2 Large dwellinghouses

A large dwellinghouse is defined in AD B1 (and is shown in Table 6.1) as having one or more floors greater than 200 m^2 in area. The guidance given in

Table 6.1 may be modified if AD B1 is used as the basis of design where the house has more than three storeys (including basements):

- It may be fitted with an L2 system as described in BS 5839 : Part 1 but the provisions in clause 16.5 regarding duration of the standby supply may be disregarded.
- Where the system is unsupervised, the standby supply should be able to automatically maintain the system in normal operation for 72 hours (but with audible and visible indication of mains failure). At the end of the 72 hours, sufficient capacity should remain to supply the maximum alarm for a minimum of 15 minutes.

6.2.3 Loft conversions

Where it is proposed to convert a loft space to habitable accommodation in a one or two-storey dwellinghouse, an automatic smoke detection and alarm system based on linked smoke alarms should be installed throughout the dwellinghouse (i.e. not just in the extended part). The installation should follow the guidance described above for the size of dwelling which will be created by the extension (i.e. either a two-storey or three-storey house). Further details of the means of escape provisions for loft conversions are given in section 6.3.9.

6.2.4 Sheltered housing

Sheltered housing usually consists of a block or group of dwellings designed specifically for persons who might require assistance (such as elderly people) where some form of assistance will be available at all times (although not necessarily on the premises). Each unit of accommodation will have its own cooking and sanitary facilities, and amenities common to all occupiers may also be provided such as communal lounges.

BS 5839 : Part 6 recommends the installation of a Grade C, Type LD3 system. The detection equipment should have a connection to a central monitoring point (or central relay station) so that the warden or supervisor is able to identify the dwelling in which the fire has occurred. It is not intended that the provisions in AD B1 or BS 5839 : Part 6 be applied to the common parts of sheltered accommodation and they do not apply to sheltered accommodation in purpose groups 2(a) Residential (institutional) or 2(b) Residential (other).

6.3 Means of escape in dwellinghouses

6.3.1 Introduction

Approved Document B1 deals with means of escape from dwellinghouses according to the height of the top storey above ground level (i.e. the ground level on the lowest side of the building). This is probably a sensible approach since storey heights can vary and means of escape through upper windows become more hazardous with increasing height. Thus, the divisions chosen are:

- houses with all floors not more than 4.5 m above ground (i.e. ground and first floor only);
- houses with one floor more than 4.5 m above ground (i.e. ground floor, first floor and second floor);
- houses with two or more floors more than 4.5 m above ground (i.e. ground floor and three or more upper floors).

Therefore, as the height of the top floor increases above ground level, the means of escape provisions become more complex and these are dealt with under separate sections below. Certain recommendations, however, are common to all dwellings and these include:

- the provision of an automatic fire detection and alarm system (see section 6.2);
- special provisions to deal with basements and inner rooms;
- windows and external doors used for escape purposes;
- balconies and flat roofs.

The guidance contained in this section is also applicable to houses in multiple occupation (HMOs) provided that there are no more than six residents. For HMOs containing greater numbers of occupants, technical guidance may be sought in DoE Circular 12/92 (see below) or Welsh Office Circular 25/92.

6.3.2 Basements and inner rooms

With certain dwelling designs (such as open-plan layouts and the provision of sleeping galleries) it is possible that a situation will be created whereby the innermost room (termed the inner room) will be put at risk by a fire occurring in the room that gives access to it (termed the access room), since escape is only possible by passing through that access room. Therefore, an inner room should only be used as:

- a kitchen, laundry or utility room;
- a dressing room;
- a bathroom, shower room or WC;
- any other room which has a suitable escape window or door (see next section for details), provided the room is in the basement, or is on the ground or first floor;
- a sleeping gallery (see section 6.3.5 for details).

Escape from a basement fire may be particularly hazardous if an internal stair has to be used, since it will be necessary to pass through a layer of smoke and hot gases. Therefore, any habitable room in a basement should have either:

- an alternative escape route via a suitable door or window; *or*
- a protected stairway leading from the basement to a final exit.

6.3.3 Windows and external doors used for escape purposes

To be suitable for escape purposes, windows and external doors should conform to the dimensions given in Fig. 6.1. Dormer windows and roof windows situated above the ground storey should comply with Fig. 6.6. Escape should be to a place of safety free from the effects of fire. Where this is to an enclosed back garden or yard from which escape may be made only by passing through other buildings, its length should be at least equivalent to the height of the dwelling (see Fig. 6.2).

6.3.4 Balconies and flat roofs

If used as an escape route, a flat roof should:

- be part of the same building from which escape is being made;
- lead to a storey exit or external escape route;
- be provided with 30 minutes' fire resistance. This applies only to the part of the flat roof forming the escape route, its supporting structure and any opening within 3 m of the escape route.

Balconies and flat roofs provided for escape purposes should be guarded in accordance with the provisions of Approved Document K[10]. (This relates to the provision of barriers at least 1100 mm high designed to prevent people falling from the escape route. The barriers should be capable of resisting at least the horizontal force given in BS 6399 : Part 1[11]).

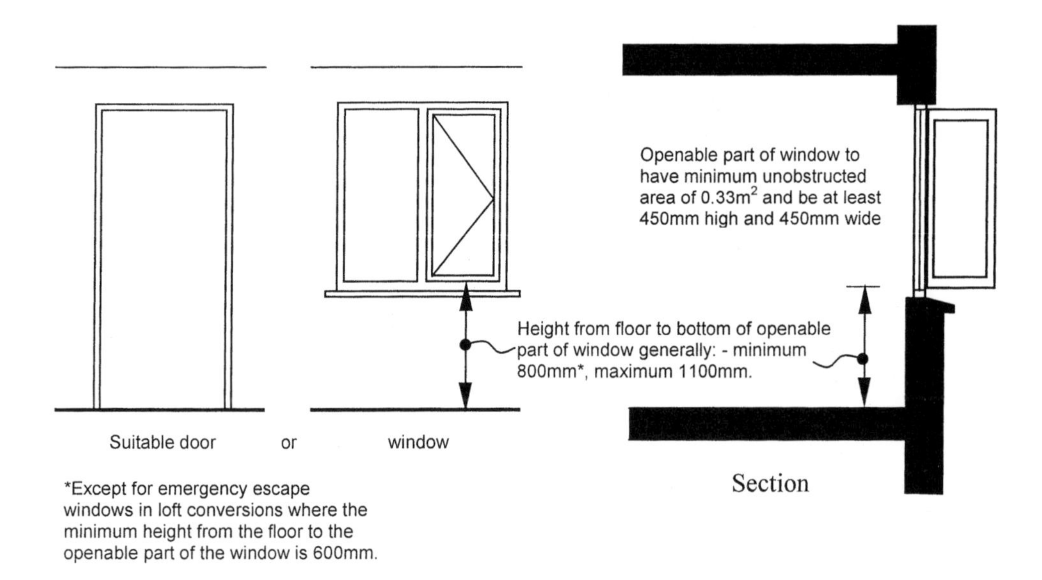

Fig. 6.1 Windows and doors for escape purposes.

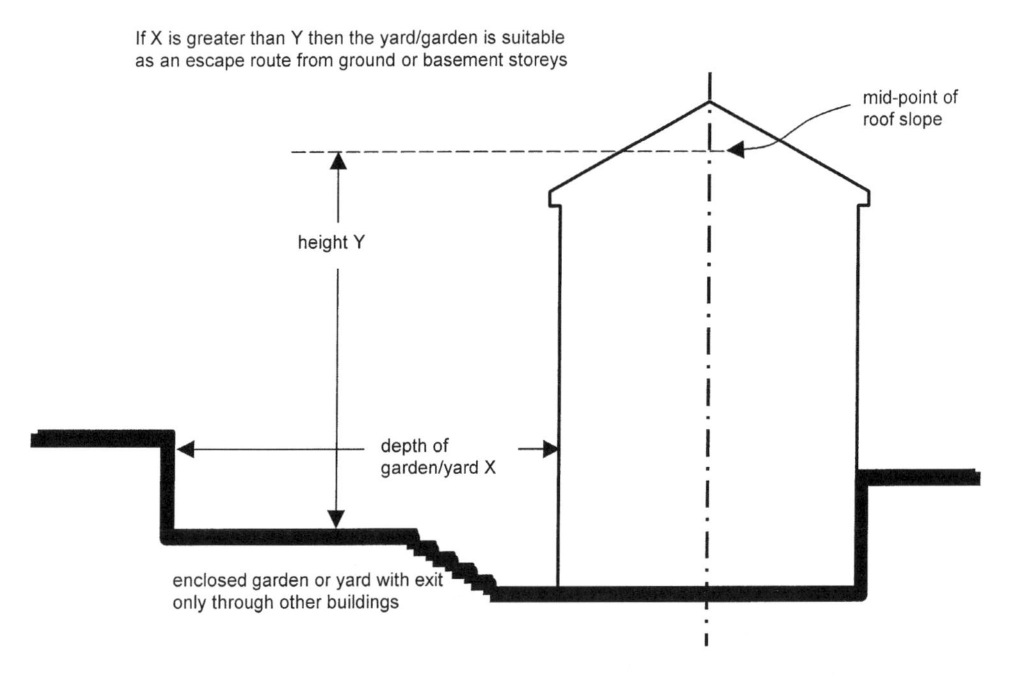

Fig. 6.2 Enclosed yard or garden suitable for escape purposes – dwellinghouses.

6.3.5 Escape from dwellinghouses with floors not more than 4.5 m above ground

For dwellings with floors not more than 4.5 m above ground, the means of escape measures outlined above need only be augmented by the following provisions:

(1) All habitable rooms (except kitchens) in the upper storey(s) of a dwellinghouse served by only one stairway, should be provided with an external door or window which complies with Fig. 6.1. A single door or window can serve two rooms provided that they each have their own access to the stairs. A communicating door should be provided between the rooms so that it is possible to gain access to the door or window without entering the stair enclosure.

(2) All habitable rooms (except kitchens) in the ground storey should either:

  - open directly onto a hall which leads to an entrance or other suitable exit, or
  - be provided with a door or window which complies with Fig. 6.1.

(3) Where the dwelling contains a sleeping gallery[12] which is not more than 4.5 m above ground level:

  - The distance between the foot of the access stair to the gallery and the door to the room which contains the gallery should not be more than 3 m.
  - An alternative exit from the gallery complying with Fig. 6.1 should be provided, if the distance from the head of the access stair to any point in the gallery exceeds 7.5 m.
  - Unless they are enclosed with fire-resisting construction, any cooking facilities within a room containing a gallery should be remote from the stair to the gallery and positioned so that they do not prejudice the escape route from the gallery.

6.3.6 Houses with one floor more than 4.5 m above ground

Houses with one floor more than 4.5 m above ground are likely to consist of ground, first and second floors. Two alternative solutions are recommended in AD B1:

(1) Provide the first and second floors with a protected stairway which should either:

  - terminate directly in a final exit; *or*

- give access to a minimum of two escape routes at ground level, separated from each other by self-closing fire doors and fire-resisting construction and each leading to a final exit;
  (these alternatives are illustrated in Fig. 6.3)

(2) Separate the top floor (and the main stairs leading to it) from the other floors by fire-resisting construction and provide it with an alternative means of escape leading to its own final exit. (A variation on this solution can be used where it is intended to convert the roof space of a two-storey dwelling, effectively making it three-storey. Further details of this are described in section 6.3.9.)

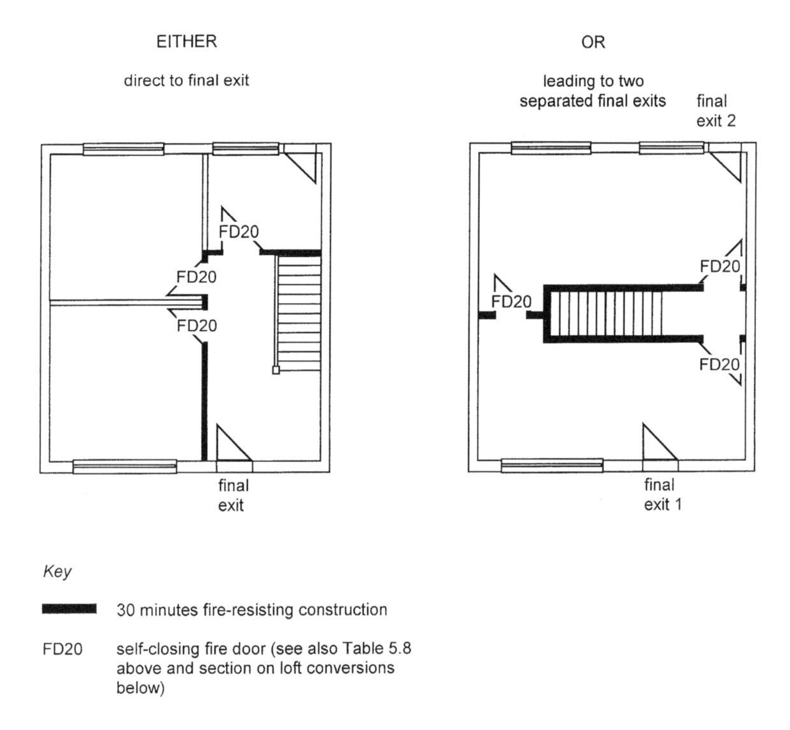

Fig. 6.3 Alternative final exit arrangements.

6.3.7 Houses with two or more floors more than 4.5 m above ground

Normally, this would include houses with three or more storeys above the ground floor. In most circumstances it will be necessary to provide an alternative means of escape for all floors which are 7.5 m or more above ground.

Where it is intended that the alternative escape route be accessed via the protected stairway (i.e. to an upper storey, see Fig. 6.4(a)) or via a landing

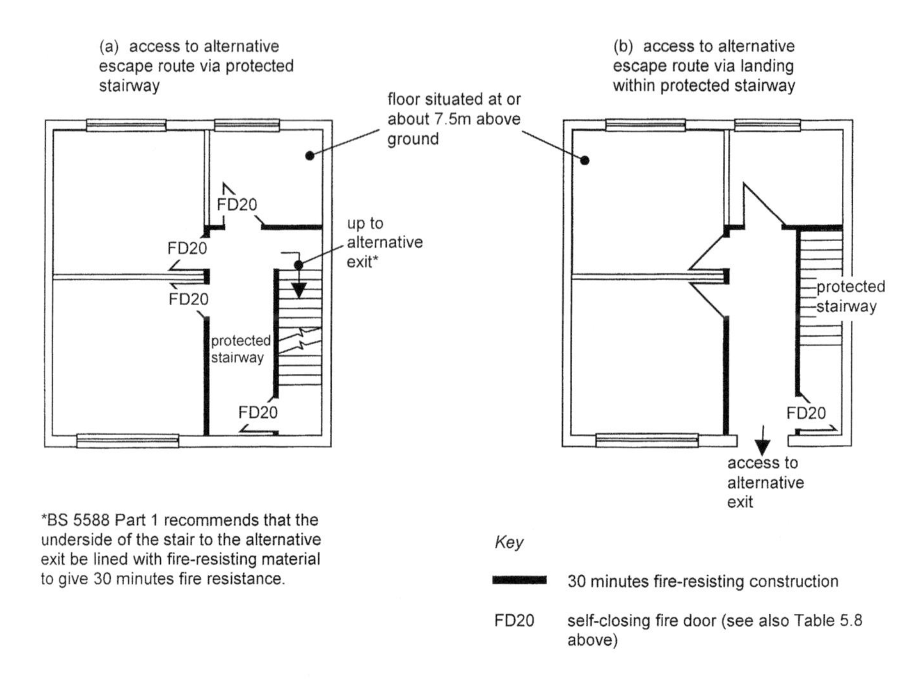

Fig. 6.4 Alternative exit arrangements in houses with more than one floor over 4.5 m above ground.

within its enclosure (i.e. on the same storey, see Fig. 6.4(b)), it is possible a fire in the lower part of the dwellinghouse might make access to the alternative route impassable. Therefore, the protected stairway at or about 7.5 m above ground level should be separated from the lower storeys by fire-resisting construction.

6.3.8 *Air circulation systems in houses with any floor more than 4.5 m above ground*

Modern houses are often fitted with air circulation systems designed for heating, energy conservation or condensation control. With such systems, there is the possibility that fire or smoke may spread (possibly by forced convection, natural convection or fire-induced convection) from the room of origin of the fire to the protected stairway. Clearly, the risk to life is greater in dwellings where there are floors at high levels so AD B1 and BS 5588 : Part 1 make the following recommendations for such buildings:

- the walls enclosing a protected stairway should not be fitted with transfer grilles;

- where ductwork passes through the enclosure to a protected stairway, it should be fitted so that all joints between the ductwork and the enclosure are fire-stopped;
- ductwork used to convey air into the protected stairway through its enclosure should be ducted back to the plant;
- grilles or registers which supply or return air should be positioned not more than 450 mm above floor level;
- any room thermostat for a ducted warm air heating system should be mounted in the living room at a height between 1370 mm and 1830 mm and its maximum setting should not be more than 27°C.
- mechanical ventilation systems which recirculate air should comply with BS 5588 : Part 9[13].

6.3.9 *Conversions to provide rooms in a roof space*

The following provisions apply when it is proposed to convert the roof space of a two-storey house to provide living accommodation. The additional floor provided is most likely to be more than 4.5 m above ground and will therefore require additional protection to the means of escape or an alternative escape route. The measures described do not apply if:

- the new floor exceeds 50 m^2 in area; *or*
- it is proposed to provide more than two habitable rooms in the new storey.

In these cases, the loft conversion will have to meet the full provisions of the Building Regulations which apply to dwellings of three or more storeys.

The recommendations for means of escape are illustrated in Figs. 6.5(a) and 6.5(b). The general principles are:

- Every doorway within the existing stairway enclosure should be fitted with a door.
- Doors to habitable rooms in the existing stair enclosure should be self-closing to prevent the movement of smoke into the means of escape. For this purpose rising butt hinges are acceptable as self-closing devices in dwellinghouses.
- New doors to habitable rooms should be fire-resisting and self-closing, but existing doors will only need to be fitted with rising butt hinges.
- The stairway in the ground and first floors should be adequately protected with fire-resisting construction and should terminate as shown in Fig. 6.3.
- Glazing (whether new or existing) in the existing stair enclosure should be fire-resisting and retained by a suitable glazing system and beads compatible

with the type of glass. This includes glazing in all doors (whether or not they need to be fire doors) except those to bathrooms or WCs. Fire-resisting glazing will usually consist of traditional annealed wired glass based on soda-lime-silica which is able to satisfy the integrity requirements of BS 476 : Part 22. There are also some unwired glass products capable of satisfying the requirements for integrity. There is no limit on the area of such glass which can be incorporated in walls and doors above 100 mm from floor level.

- The new storey should be served by a stair which is located as shown in Figs. 6.5(a) and 6.5(b) and complies with Approved Document K. This could be an alternating tread stair.
- The new storey should be separated from the remainder of the dwelling by fire-resisting construction. Any openings should be protected by self-closing fire doors situated as shown in Figs 6.5(a) and 6.5(b).
- Windows and doors provided for emergency escape in the basement, ground and first floors of the existing dwelling provide a means of self-rescue. At higher levels, it is acceptable in loft conversions for escape to

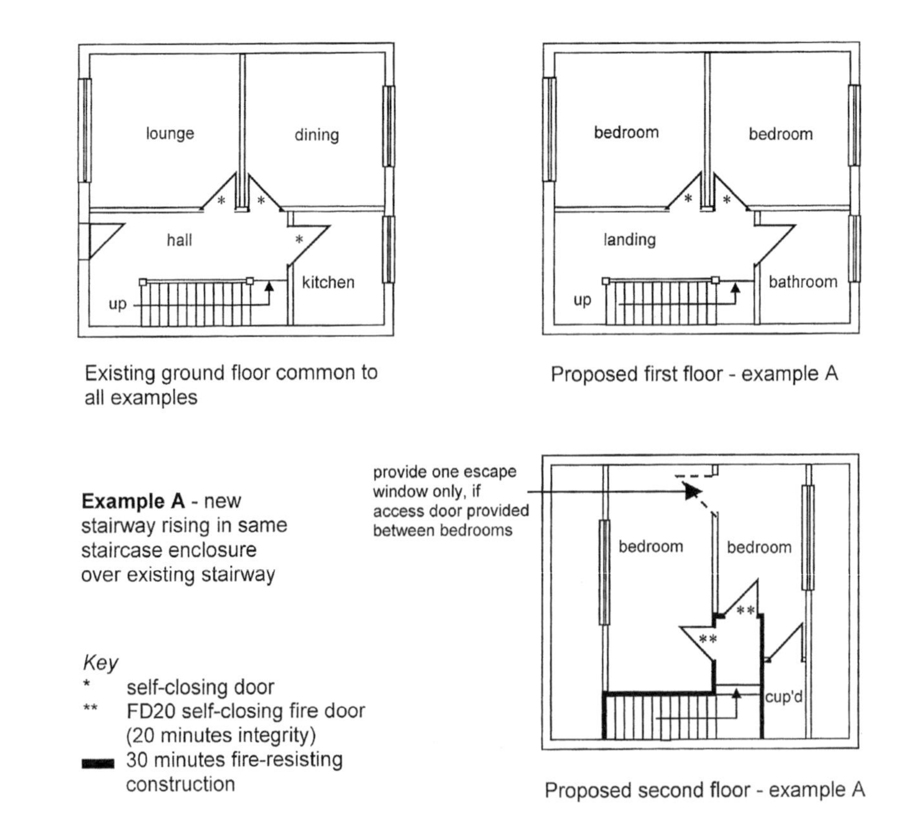

Fig. 6.5(a) Loft conversion to existing two-storey dwellinghouse.

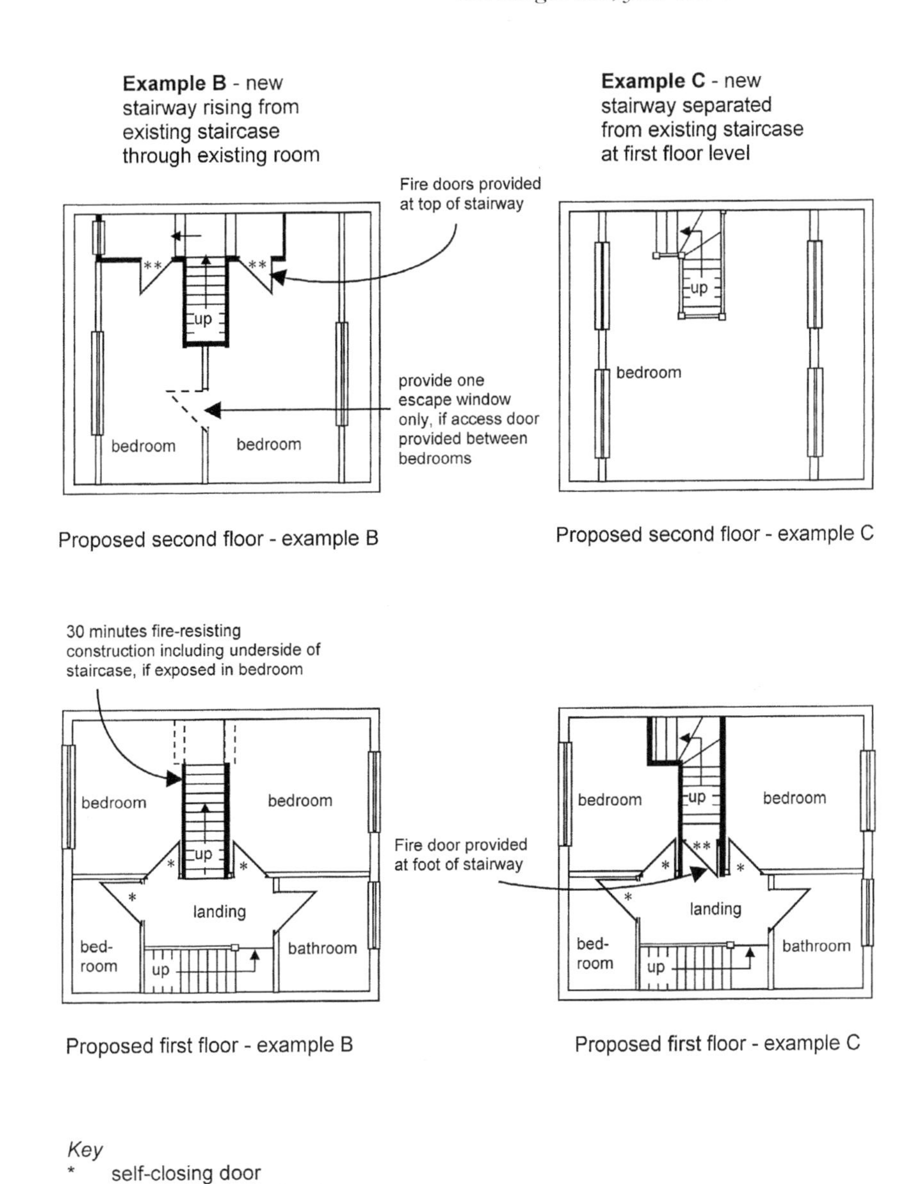

Fig. 6.5(b) Loft conversion to existing two-storey dwellinghouse.

depend on a ladder being set up to a suitable escape window, or to a roof terrace, which is accessible by a door from the loft conversion. Any escape window should be large enough and suitably positioned to allow escape by means of a ladder to the ground, see Fig. 6.6.

- There should be suitable means of pedestrian access to the place at which the ladder would be set, to allow fire service personnel to carry a ladder from their vehicle (although it should not be assumed that only the fire service will effect a rescue). A ladder fixed to the slope of the roof is not recommended due to the danger inherent in using such ladders.
- Where the loft conversion consists of two rooms a single window can serve both rooms provided that they each have their own access to the stairs. A communicating door should be provided between the rooms so that it is possible to gain access to the window without entering the stair enclosure.
- Escape across the roof of a ground storey extension is acceptable provided that the roof complies with the recommendations listed in section 6.3.4. Additionally, when it is proposed to erect an extension to a dwelling, the effect of the extension on the ability to escape from windows in other parts of the dwelling (especially from a loft conversion) should be carefully considered. This is particularly true when erecting a conservatory with a glazed roof.

The examples shown in Figs 6.5(a) and 6.5(b) are typical illustrations of the principles outlined above for one particular form of dwellinghouse. They should not be taken to be definitive in any sense.

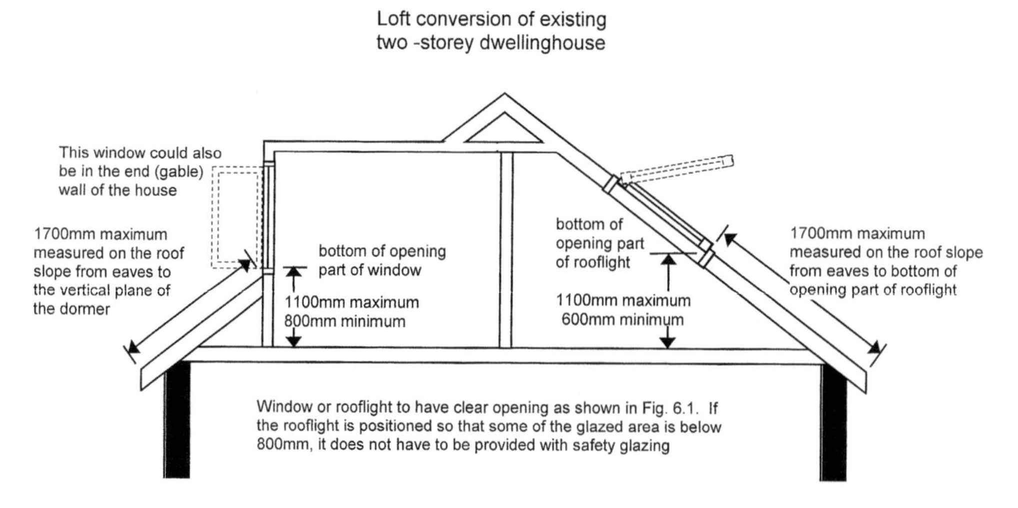

Fig. 6.6 Window positions for emergency escape dormers and rooflights in loft conversions.

Flats and maisonettes

Approved Document B1 deals with a limited range of common designs for flats and maisonettes. Where less common arrangements are desired (for example, where flats are entered above or below accommodation level, or flats contain sleeping galleries) the principles described in this chapter may still be applied or reference can be made to BS 5588 : Part 1: 1990, clauses 9 and 10.

The means of warning and escape recommendations listed above for dwellinghouses which consist of basement, ground and first floors only, apply equally to flats and maisonettes situated at these levels. At higher levels, escape through upper windows becomes more hazardous and more complex provisions are necessary especially in maisonettes where internal stairs will need protection.

In addition to the general assumptions stated in Chapter 4 for all building types, the following assumptions are made when considering the means of warning and escape from flats and maisonettes:

- fires generally originate in the dwelling;
- rescue by ladders is not considered suitable;
- fire spread beyond the dwelling of origin is unlikely due to the compartmentation recommendations in Approved Document B3, therefore simultaneous evacuation of the building should be unnecessary;
- fires which occur in common areas are unlikely to spread beyond the immediate vicinity of the outbreak due to the materials and construction used there.

6.4 Fire alarm and detection systems in flats and maisonettes

The principles for the provision of fire alarm and detection systems in flats and maisonettes are the same as those which apply to dwellinghouses (see section 6.2), with the following additions:

- there is no need to apply the provisions to the common parts of blocks of flats;
- the systems in individual flats do not need to be interconnected;
- a maisonette (i.e. a 'flat' in which the accommodation is contained on more than one level and which may be entered from either the higher or lower level) should be treated in the same way as a house with more than one storey.

6.5 Means of escape in flats and maisonettes

There are two main components to the means of escape from flats and maisonettes:

- escape from within the dwelling itself; *and*
- escape from the dwelling to the final exit from the building (usually along a common escape route).

The following sections consider these two components.

6.5.1 Means of escape from within flats and maisonettes

The provisions relating to inner rooms, basements, balconies and flat roofs in dwellings (see sections 6.3.2 to 6.3.4) also apply to the inner parts of flats and maisonettes.

Flats and maisonettes with floors not more than 4.5 m above ground

Generally, where the floor of a flat or maisonette is not more than 4.5 m above ground it should be planned so that any habitable room in a ground or upper storey is provided with a suitable escape window or door, as shown in Fig. 6.1 above. A single door or window in an upper storey can serve two rooms provided that they each have their own access to the stair enclosure or entrance hall. A communicating door should be provided between the rooms so that it is possible to gain access to the door or window without entering the stair enclosure or entrance hall.

Alternatively, upper floors can be designed to follow the remainder of the guidance described below, in which case escape windows in upper storeys will be unnecessary.

Flats and maisonettes with floors more than 4.5 m above ground

Provided that the restrictions on inner rooms are observed, three possible solutions to the internal planning of flats are given in AD B1:

(1) All the habitable rooms in the flat are arranged to have direct access to a protected entrance hall. The maximum distance from the entrance door to the door of a habitable room should not exceed 9 m. (See Fig. 6.7(a)).

(2) In flats where a protected entrance hall is not provided, the 9 m distance referred to in (1) should be taken as the furthest distance from any point in

the flat to the entrance door. Cooking facilities should be remote from the entrance door and positioned so that they do not prejudice the escape route from any point in the flat. (See Fig. 6.7(b)).

(3) Provide an alternative exit from the flat. In this case, the internal planning will be more flexible. An example of a typical flat plan where an alternative exit is provided, but not all the habitable rooms have direct access to the entrance hall, is shown in Fig. 6.7(c).

Where flats and maisonettes with floors more than 4.5 m above ground are fitted with air circulation systems designed for heating, energy conservation or condensation control there is the possibility that fire or smoke may spread from the room of origin of the fire to the protected entrance hall or landing. Clearly, the risk to life is greater where there are floors at high levels so AD B1 and BS 5588 : Part 1 make the following recommendations for such buildings:

- Any wall, door, floor or ceiling enclosing a protected entrance hall of a dwelling or protected stairway and landing of a maisonette should not be fitted with transfer grilles.
- Where ductwork passes through the enclosure to a protected entrance hall or protected stairway and landing, it should be fitted so that all joints between the ductwork and the enclosure are fire-stopped.
- Ductwork used to convey air into the protected entrance hall of a dwelling or protected stairway and landing of a maisonette through the enclosure of the protected hall or stairway, should be ducted back to the plant.
- Grilles or registers which supply or return air should be positioned not more than 450 mm above floor level.
- Any room thermostat for a ducted warm air heating system should be mounted in an area from which air is drawn directly to the heating unit at a height between 1370 mm and 1830 mm and its maximum setting should not be more than 27°C.
- Mechanical ventilation systems which recirculate air should comply with BS 5588 : Part 9.

Maisonette with independent external entrance at ground level

A maisonette of this type is similar to a dwellinghouse and should have a means of escape which complies with the recommendations for dwellings described in section 6.3, depending on the height of the top storey above ground.

Maisonette with floor more than 4.5 m above ground and no external entrance at ground level

Two internal planning arrangements are described in AD B1 for maisonettes of this type. These are illustrated in Fig. 6.8(a) and (b), as follows:

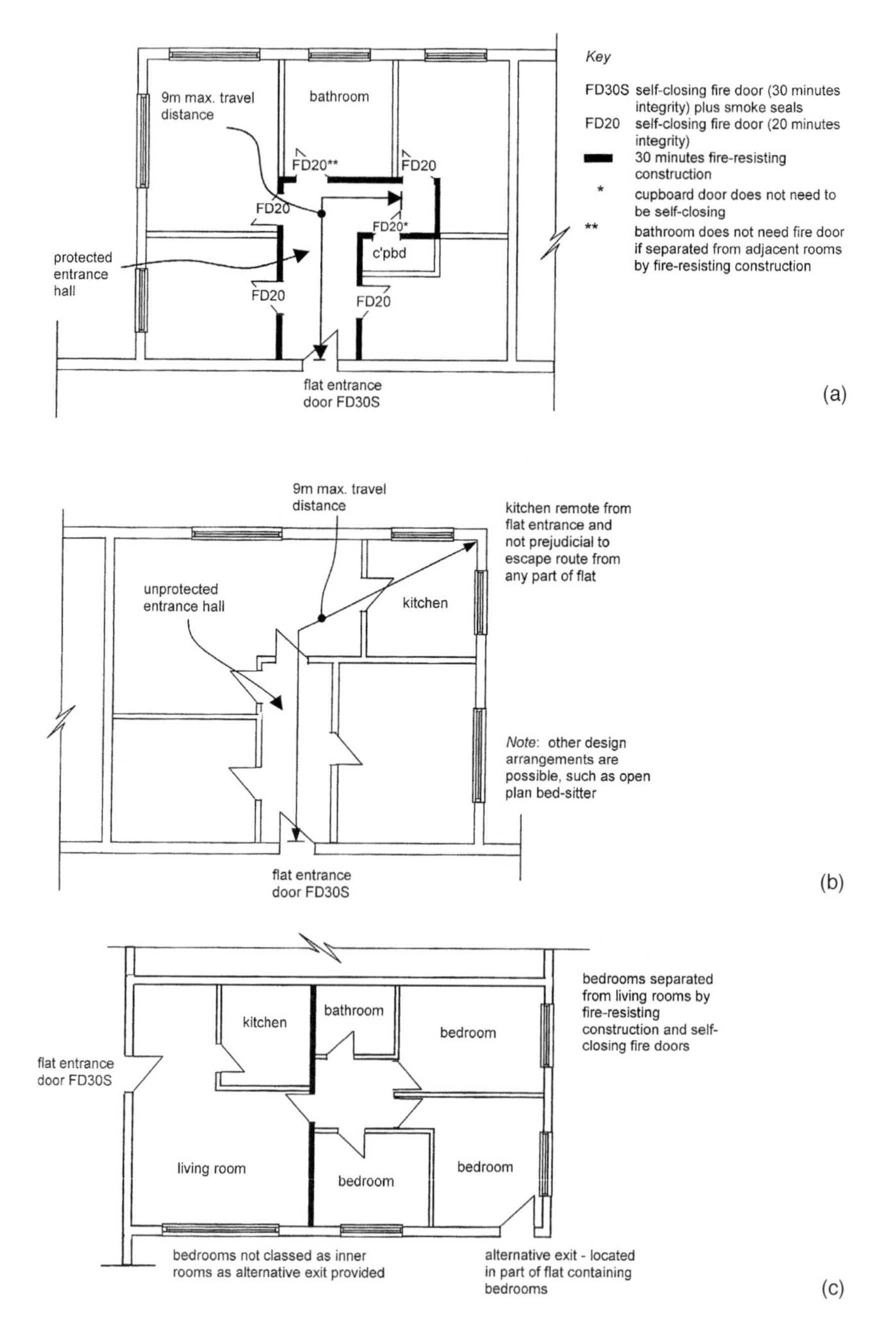

Fig. 6.7 Examples of alternative internal layouts to flats. (a) Flat with protected entrance hall; (b) Flat with unprotected entrance hall and limited travel distance; (c) Flat with alternative exit but no direct access for all habitable rooms to entrance hall.

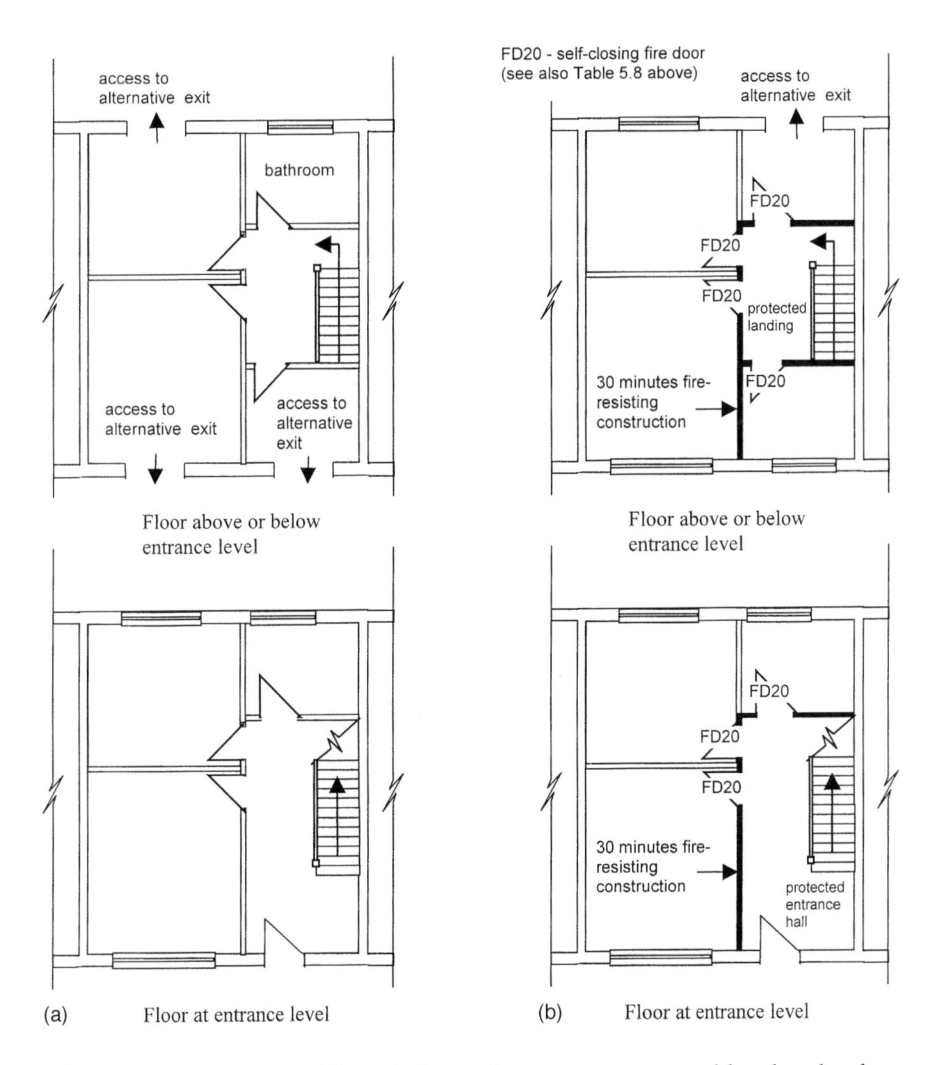

Fig. 6.8 Maisonette with no independent access at ground level and at least one storey more than 4.5 m above ground. (a) Maisonette with alternative exits from habitable rooms (except entrance level); (b) Maisonette with protected landing and entrance hall.

- provide each habitable room which is not on the entrance floor with an alternative exit; *or*
- provide a protected entrance hall and/or landing entered directly from all the habitable rooms on that floor; additionally, one alternative exit should be provided on each floor which is not the entrance floor.

6.5.2 Alternative exits from flats and maisonettes

That part of the means of escape from the entrance door of a flat or maisonette to a final exit from the building is often by way of a route which is common to all dwellings in the block. Much of the information covering vertical and horizontal escape in Chapter 4 is relevant to common escape routes and should be read in conjunction with the guidance described below, although none of this is applicable to such buildings where the top floor is not more than 4.5 m above ground level.

Reference has been made above to the provision of alternative exits in certain planning arrangements for flats and maisonettes. Alternative exits will only be effective if they are remote from the main entrance door to the dwelling and lead to a common stair or final exit by means of:

- a door to an access corridor, access lobby or common balcony; *or*
- an internal private stairway leading to an access corridor, access lobby or common balcony on another level; *or*
- a door onto a common stair; *or*
- a door to an external stair; *or*
- a door to an escape route over a flat roof.

6.5.3 Means of escape in the common parts of flats and maisonettes

In general, flats and maisonettes should have access to an alternative means of escape. In this way, it will be possible to escape from a fire in a neighbouring flat by walking away from it. It is not always possible to provide alternative escape routes (which means providing two or more staircases) in all buildings containing flats and maisonettes, therefore single staircase buildings are permissible in certain circumstances. Typical examples of single and multi-stair buildings are described below.

Flats and maisonettes with single common stairs

In larger buildings, it will be necessary to separate the entrance to each dwelling from the common stair by a protected lobby or common corridor. The maximum distance from any entrance door to the common stair or protected lobby should not exceed 7.5 m. (See Fig. 6.9(a) and (b).)

These recommendations may be modified for smaller buildings where:

- the building consists of a ground storey and no more than three other storeys above the ground storey;

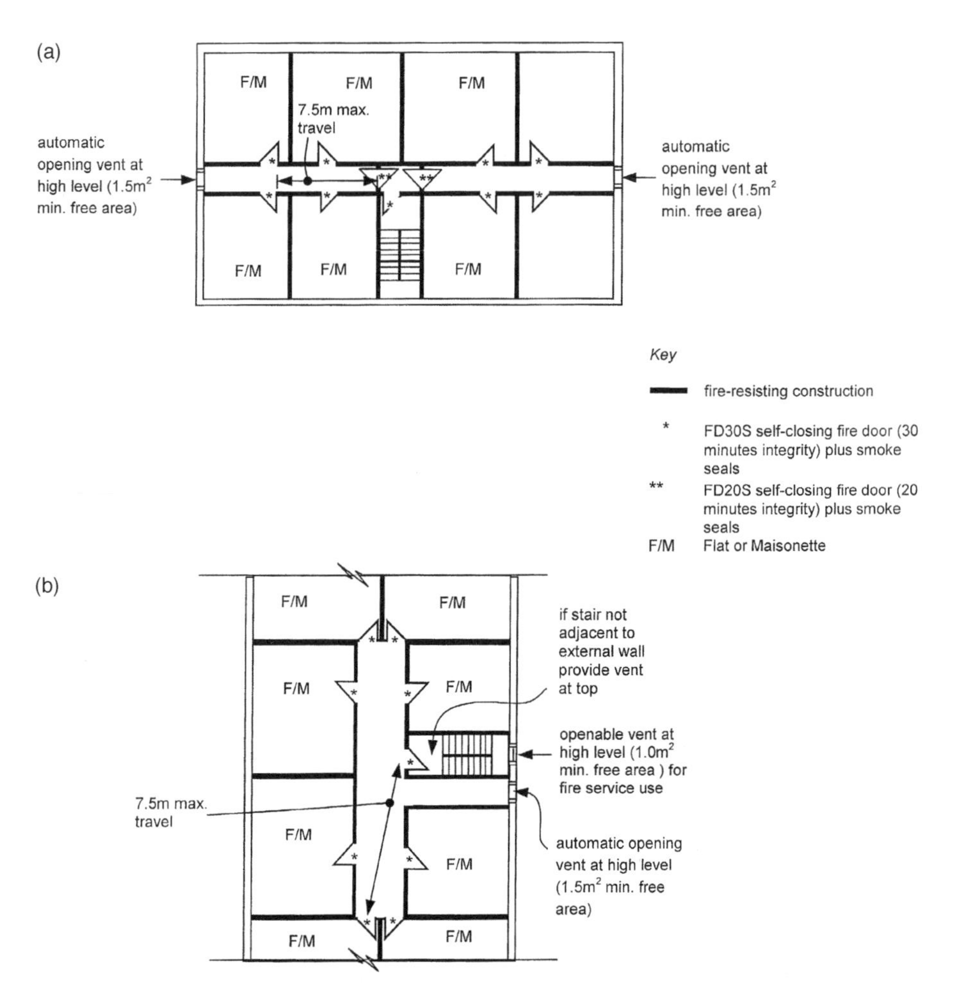

Fig. 6.9 Flats and maisonettes with single common stairs. (a) Corridor access flats and maisonettes – single common stair; (b) Lobby access flats and maisonettes – single common stair.

- the top floor does not exceed 11 m above ground level;
- the stair does not connect to a covered car park unless it is open-sided;
- the stair does not also serve ancillary accommodation (see end of section 4.2), although this does not apply to ancillary accommodation:
  - in any storey which does not contain dwellings; *and*
  - which is separated from the stair by a ventilated protected lobby or ventilated protected corridor (i.e. provide permanent ventilation of not less than 0.4 m^2 or a mechanical smoke control system to prevent the ingress of smoke).

The modified recommendations are illustrated in Fig. 6.10(a). The maximum distance from the dwelling entrance door to the stair entrance should be reduced to 4.5 m. If the intervening lobby is provided with an automatic opening vent this distance may be increased to 7.5 m. Where the building contains only two flats per floor further simplifications as shown in Fig. 6.10(b) and (c) are possible.

Flats and maisonettes with more than one common stair

Where escape is possible in two directions from the dwelling entrance door, the maximum escape distance to a storey exit may be increased to 30 m. Furthermore, if all the dwellings on a storey have independent alternative means of escape, the maximum travel distance of 30 m does not apply. (There is still, however, the need to comply with the fire service access recommendations in Approved Document B5, where vehicle access for a pump appliance should be within 45 m of every dwelling entrance door.)

In buildings of this type it is possible to have a dead-end situation where the stairs are not located at the extremities of each storey. This is permissible provided that the dead-end portions of the corridor are fitted with automatic opening vents and the dwelling entrance doors are within 7.5 m of the common stair entrance.

Typical details of flats and maisonettes with more than one common stair are shown in Fig. 6.11.

Flats and maisonettes with balcony or deck approach

This is a fairly common arrangement whereby all dwellings are accessed by a continuous open balcony or deck on one or both sides of the block. AD B1 refers the reader to clause 13 of BS 5588 : Part 1, on which the following notes are based.

The principal risk in such arrangements is smoke-logging of the balconies or decks. This is less likely to occur when the balconies are relatively narrow, therefore the only considerations necessary are:

- that vehicle access for a fire service pump appliance is within 45 m of every dwelling entrance door, and all parts of the building are within 60 m of a fire main;
- in the case of single stair buildings, that persons wishing to escape past the dwelling on fire can do so safely. This is usually achieved by ensuring that the external part of each dwelling facing the balcony is protected by 30 minutes' fire-resisting construction for a distance of 1100 mm from the balcony floor level.

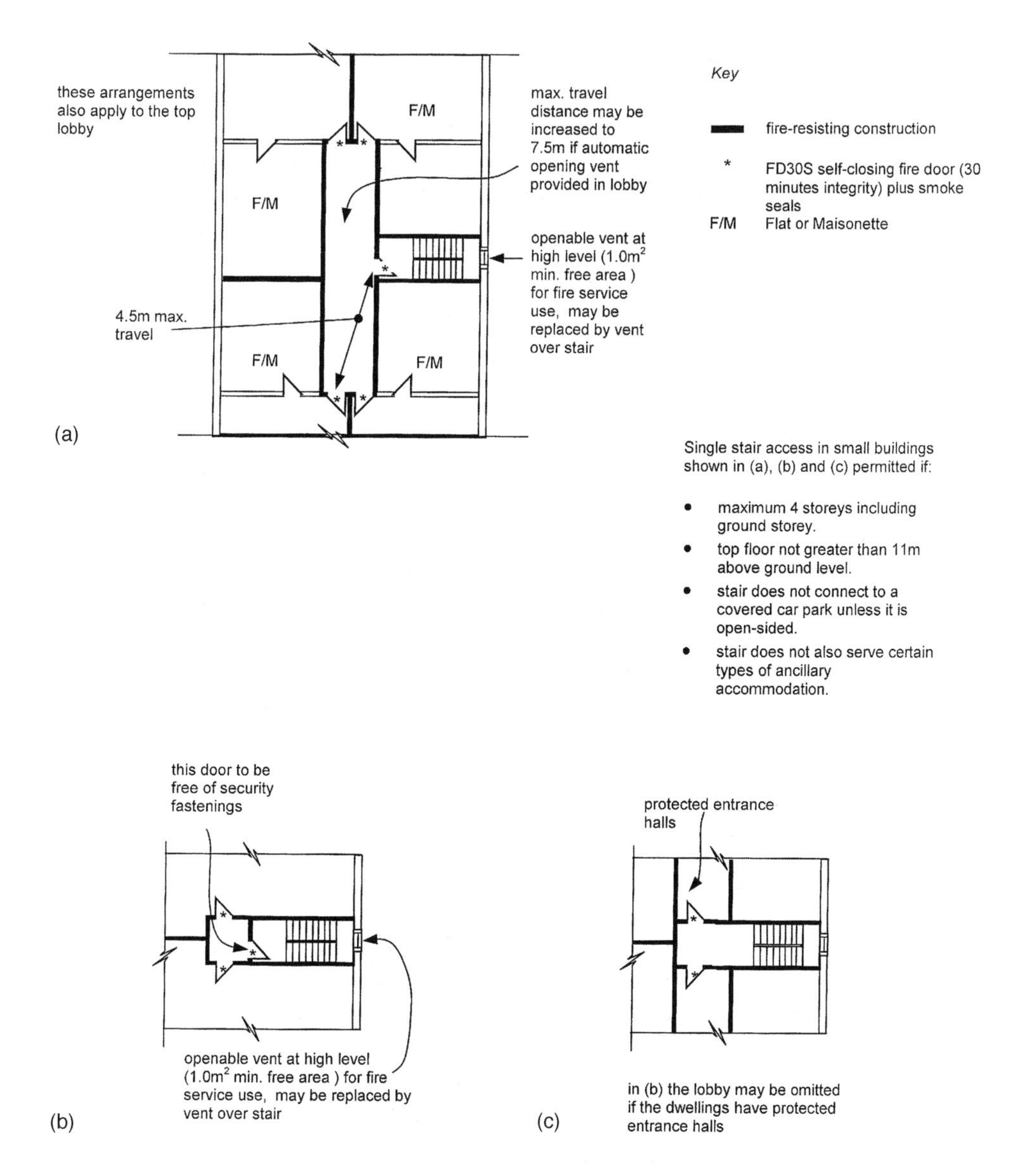

Fig. 6.10 Flats and maisonettes – small single stair buildings. (a) Common escape route for flats and maisonettes in small single stair building; (b) Small single stair building – maximum two dwellings per floor; (c) Small single stair building – maximum two dwellings per floor with protected entrance halls.

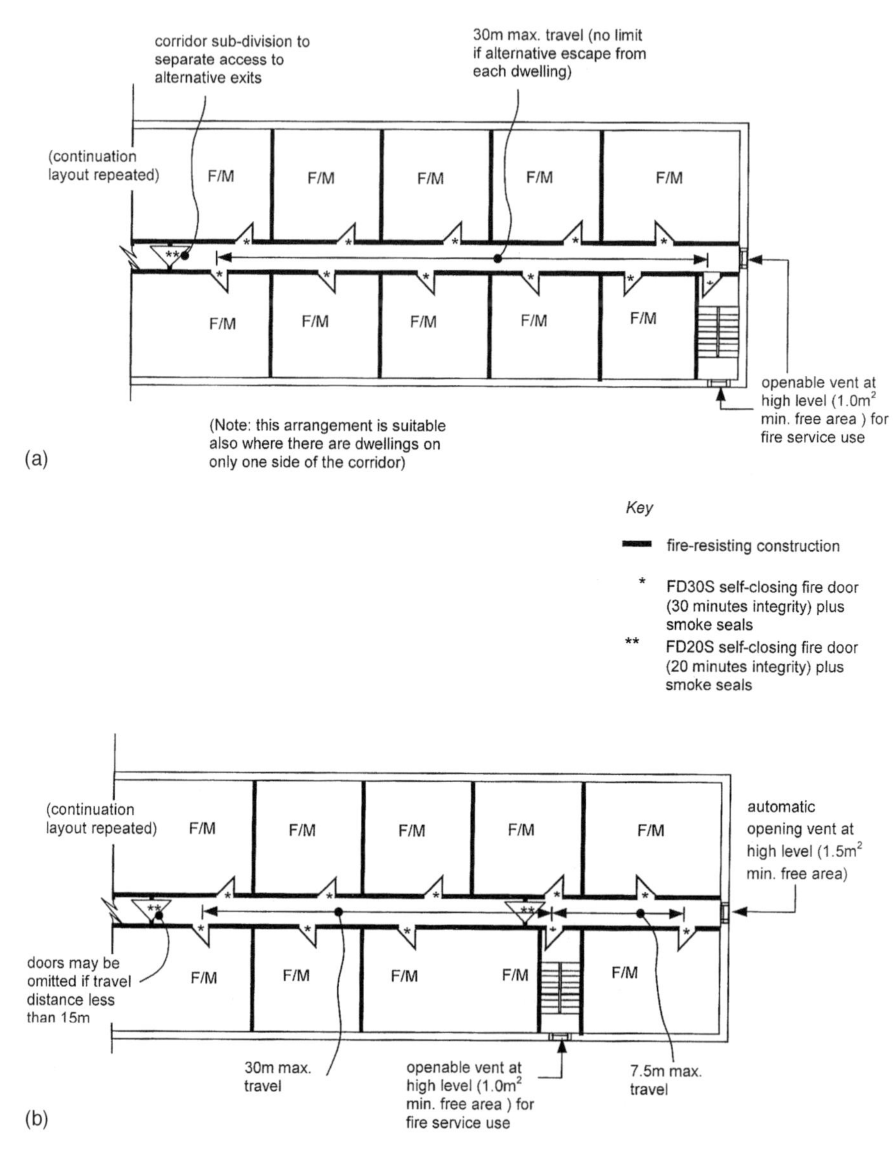

Fig. 6.11 Flats and maisonettes with more than one common stair. (a) Corridor access without dead-end; (b) Corridor access with dead-end.

Smoke-logging is more likely to occur with the adoption of wider balconies or a deck approach. The provision of downstands from the soffit above a deck or balcony at right angles to the face of the building can reduce the possibility of smoke from any dwelling on fire spreading laterally along the deck. This would also reduce the chances of smoke logging on the decks above. Therefore, where the soffit above a deck or a balcony has a width of 2 m or more:

- it should be designed with downstands placed at 90° to the face of the building (on the line of separation between individual dwellings); and
- the down-stand should project 300 mm to 600 mm below any other beam or downstand parallel to the face of the building.

There is a risk that occupants of dwellings with wider balconies or deck approaches will use this opportunity of greater depth to erect 'external' stores or other fire risks. Therefore, no store or other fire risk should be erected externally on the balcony or deck.

Examples of common escape routes for dwellings with balcony or deck access are shown in Fig. 6.12.

Additional provisions for common escape routes

Common escape routes should be planned so that they are not put at risk by a fire in any of the dwellings or in any stores or ancillary accommodation. The following recommendations are designed to provide additional protection to these routes:

- It should not be necessary to pass through one stairway enclosure to reach another. Where this is unavoidable a protected lobby should be provided to the stairway. This lobby may be passed through in order to reach the other stair.
- Common corridors should be designed as protected corridors and should be constructed to be fire-resisting.
- The wall between each dwelling and the common corridor should be a compartment wall.
- A common corridor connecting two or more storey exits should be sub-divided with a self-closing fire door and/or fire-resistant screen, positioned so that smoke will not affect access to more than one storey exit.
- A dead-end section of a common corridor should be separated in a similar manner from the rest of the corridor.
- Protected lobbies and corridors should not contain any stores, refuse chutes, refuse storage areas or other ancillary accommodation. See also Chapter 5, page 115, above for further details of the provision of refuse chutes and stores.

Ventilation of common escape routes

Although precautions can be taken to prevent the ingress of smoke onto common corridors and lobbies, it is almost inevitable that there will be some leakage since a flat entrance door must be opened in order that the occupants

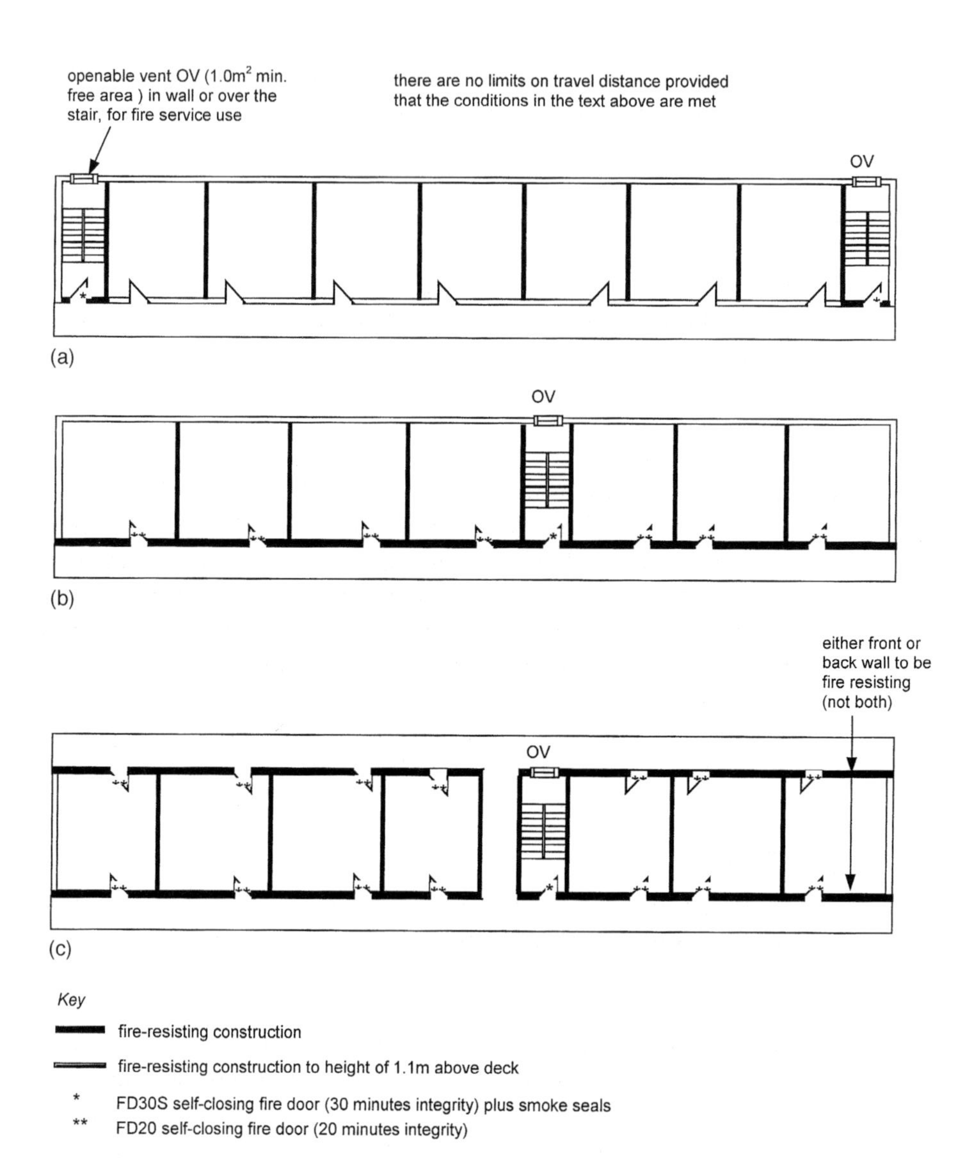

Fig. 6.12 Flats and maisonettes with balcony/deck approaches. (a) Multi-stair building, no alternative exits from dwellings; (b) Single stair building, no alternative exits from dwellings; (c) Single stair building with alternative exit from each dwelling.

can escape. Provisions for ventilation of the common areas (which also provide some protection for common stairs) are therefore vital and may be summarised as follows:

- Subject to the variations shown in Fig. 6.9(a) and (b), common corridors or lobbies in larger, single-stair buildings should be provided with automatic opening ventilators, triggered by automatic smoke detection located in the space to be ventilated. These should be positioned as shown in the figure, should have a free area of at least 1.5 m^2 and should be fitted with a manual override.
- Small single-stair buildings should conform to the guidance shown in Fig. 6.10(a), (b) and (c).
- Common corridors in multi-stair buildings should extend at both ends to the external face of the building where openable ventilators, or automatic opening ventilators, should be fitted for fire service use. They should have a free area of 1.0 m^2 at each end of the corridor (see Fig. 6.11).
- It is possible to protect escape stairways, corridors and lobbies by means of smoke control systems employing pressurisation. These systems should comply with BS 5588 : Part 4[14] : 1998. Where these are provided the cross corridor fire doors and the openable and automatic opening ventilators referred to above should be omitted.

Escape routes across flat roofs

Where more than one escape route exists from a storey or part of a building, one of those routes may be across a flat roof if the following conditions are observed:

- The flat roof should be part of the same building.
- The escape route over the flat roof should lead to a storey exit or external escape route.
- The roof and its structure forming the escape route should be fire-resisting.
- Any opening within 3 m of the route should be fire-resisting.
- The route should be adequately defined and guarded in accordance with Approved Document K. (This relates to the provision of barriers at least 1100 mm high designed to prevent people falling from the escape route. The barriers should be capable of resisting at least the horizontal force given in BS 6399 : Part 1).

Provision of common stairs in flats and maisonettes

Stairs which are used for escape purposes should provide a reasonable degree of safety during evacuation of a building. Since they may also form a potential route for fire spread from floor to floor there are recommendations contained in Approved Document B3 which are designed to prevent this. Stairs may also be used for fire-fighting purposes. In this case reference should be made to the recommendations contained in Approved Document B5 (see Chapter 5, Table 5.4). The following recommendations are specifically for means of escape purposes:

- Each common stair should be situated in a fire-resisting enclosure with the appropriate level of fire resistance taken from Tables Al and A2 of Appendix A of Approved Document B.
- Each protected stair should discharge either:
  - direct to a final exit; *or*
  - by means of a protected exit passageway to a final exit.
- If two protected stairways or protected exit passageways leading to different final exits are adjacent, they should be separated by an imperforate enclosure.
- A protected stairway should not be used for anything else apart from a lift well or electricity meters.
- Openings in the external walls of protected stairways should be protected from fire in other parts of the building if they are situated where they might be at risk. (See end of section 5.2.8 and Fig. 5.8 for details of protection measures.)
- A stair of acceptable width for everyday use will also be sufficient for escape purposes (BS 5588 : Part 1 recommends a minimum width of 1 m); however, if the stair is also a fire-fighting stair this should be increased to 1.1 m.
- Basement stairs will need to comply with special measures (see end of section 5.2.6).
- Gas service pipes and meters should only be installed in protected stairways if the installation complies with the requirements for installation and connection set out in the Pipelines Safety Regulations[15] 1996 and the Gas Safety (Installation and Use) Regulations[16] 1998.
- A common stair which forms part of the only escape route from a flat or maisonette should not also serve any fire risk area such as a covered car park, boiler room, fuel storage space or other similar ancillary accommodation on the same storey as that dwelling (but see the exceptions to this mentioned above).
- Where more than one common stair is provided as an escape from a dwelling it is permitted to serve ancillary accommodation provided that it is separated from that accommodation by a protected lobby or protected corridor.
- Where any stair serves an enclosed car park or place of special fire hazard

(see Table 5.2) it should be separated from that accommodation by a ventilated lobby or ventilated corridor (i.e. provide permanent ventilation of not less than 0.4 m^2 or a mechanical smoke control system to prevent the ingress of smoke).

External access and escape stairs

Where the building (or any part of it) is permitted to be served by a single access stair, that stair may be placed externally if it serves a floor which is not more than 6 m above ground level and it complies with the provisions listed in (2) to (6) below (see Fig. 6.13(a)).

Where there is more than one escape route available from a storey or part of a building some of the escape routes may be by way of an external escape stair if there is at least one internal escape stair serving every part of each storey (except plant areas) and if the following provisions (illustrated in Fig. 6.13) can be met:

(1) The stair should not serve any floors which are more than 6 m above the ground or a roof or podium. The roof or podium should itself be served by an independent protected stair.

(2) If the stair is more than 6 m in vertical extent, it should be sufficiently protected from adverse weather. This does not necessarily mean that full enclosure will be necessary. The stair may be located so that protection may be obtained from the building itself. In deciding on the degree of protection it is necessary to consider the height of the stair, the familiarity of the occupants with the building, and the likelihood of the stair becoming impassable as a consequence of adverse weather conditions.

(3) Any part of the building (including windows and doors, etc.) which is within 1.8 m of the escape route from the stair to a place of safety should be protected with fire-resisting construction. This does not apply if there is a choice of routes from the foot of the stair, thereby enabling the people escaping to avoid the effects of fire in the adjoining building. Additionally, any part of an external wall which is within 1.8 m of an external escape route (other than a stair) should be of fire-resisting construction up to a height of 1.1 m from the paving level of the route.

(4) All the doors which lead onto the stair should be fire-resisting and self-closing. This does not apply to the only exit door to the landing at the head of a stair which leads downward.

(5) Any part of the external envelope of the building which is within 1.8 m of (and 9 m vertically below) the flights and landings of the stair, should be of fire-resisting construction. This 1.8 m dimension may be reduced to 1.1 m above the top landing level provided that this is not the top of a stair up from basement level to ground.

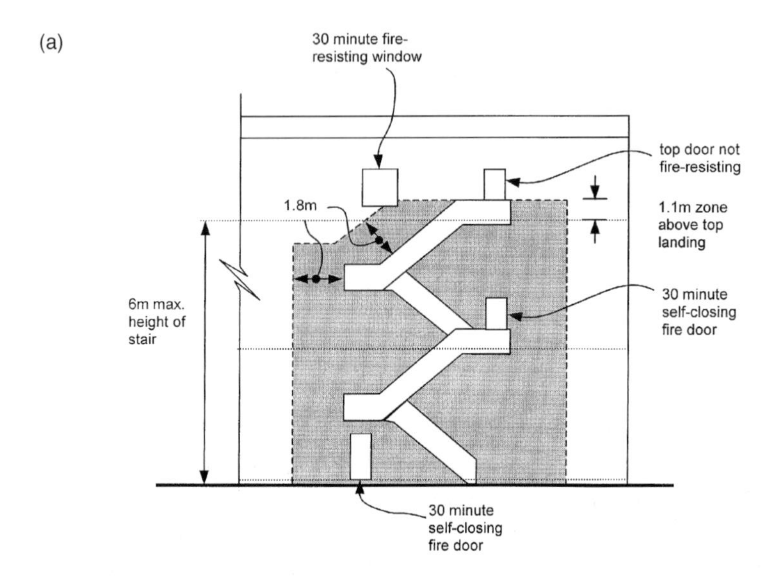

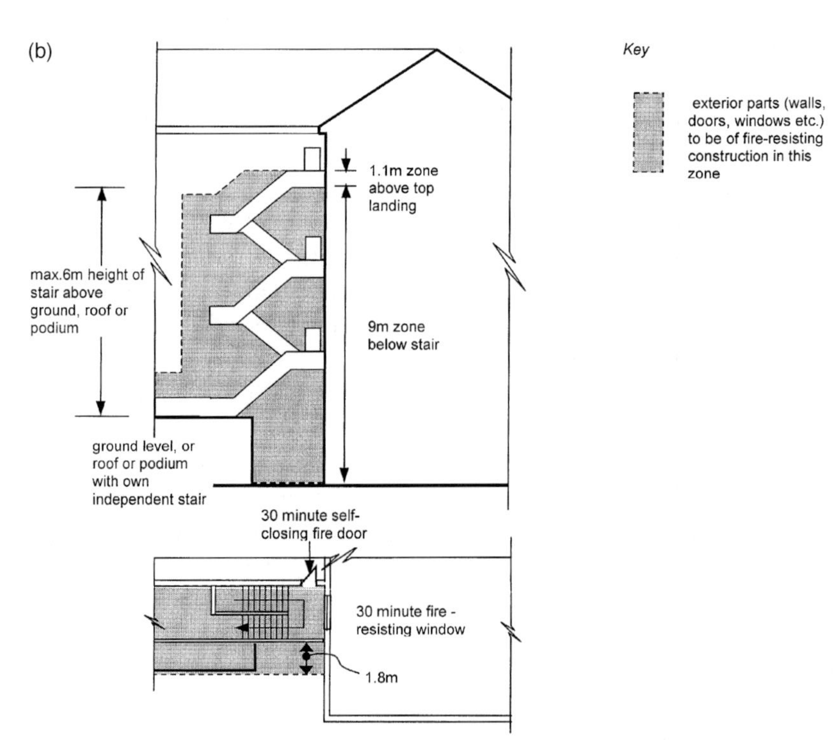

Fig. 6.13 External escape stairs to flats and maisonettes. (a) Flats and maisonettes with external single access stair; (b) External escape stair to flats and maisonettes with alternative internal escape stairs.

(6) Any glazing which is contained within the fire-resisting areas mentioned above should also be fire-resisting in terms of maintaining its integrity in a fire, and be fixed shut. (For example, Georgian wired glass is adequate; it does not also have to meet the requirements for insulation).

Stairs to dwellings in mixed use buildings

Many buildings consist of a mix of dwellings (i.e. flats and maisonettes) and other uses. Sometimes the dwellings are ancillary to the main use (such as a caretaker's flat in an office block), and sometimes they form a distinct separate use (as in the case of shops with flats over). Clearly, the degree of separation of the uses for means of escape purposes will depend on the height of the building and the extent to which the uses are interdependent.

Where a building has no more than three storeys above the ground storey, the stairs may serve both non-residential and dwelling uses, with the proviso that each occupancy is separated from the stairs by protected lobbies at all levels.

In larger buildings where there are more than three storeys above the ground storey, stairs may serve both the dwellings and the other occupancies if:

- The dwelling is ancillary to the main use of the building and is provided with an independent alternative escape route.
- The stair is separated from other parts of the building at lower storey levels by protected lobbies at those levels and has the same standard of fire resistance as that required by Approved Document B for the rest of the building (including any additional provisions if it is a fire-fighting stair).
- Any automatic fire detection and alarm system fitted in the main part of the building is extended to the flat.
- Any security (usually installed for the benefit of the non-dwelling use) does not prevent escape at all material times.

Where fuels, such as petrol and liquid petroleum gas are stored, additional measures (such as an increase in the fire resistance period for the structure between the storage area and the dwelling) may be necessary.

6.6 References

1 BS 5839 Fire detection and alarm systems for buildings, Part 1 : 1988 Code of practice for system design, installation and servicing.
2 BS 5839 Fire detection and alarm systems for buildings, Part 6 : 1995 Code of practice for the design and installation of fire detection and alarm systems in dwellings.

3 Firecode. HTM 88. Guide to fire precautions in NHS housing in the community for mentally handicapped (or mentally ill) people. (DHSS) HMSO, 1986.
4 DoE Circular 12/92 Houses in multiple occupation. Guidance to local housing authorities on standards of fitness under section 352 of the Housing Act 1985.
5 Welsh Office Circular 25/92 Local Government and Housing Act 1989. Houses in multiple occupation: standards of fitness.
6 BS 5839 Fire detection and alarm systems for buildings, Part 4 : 1988 Specification for control and indicating equipment.
7 BS 5446 Components of automatic fire alarm systems for residential premises, Part 1 : 1990 Specification for self-contained smoke alarms and point-type detectors. This code deals with smoke alarms based on ionization chamber smoke detectors and optical (photo-electric) smoke detectors. Each type of detector responds differently to smouldering and fast flaming fires. Therefore, optical smoke alarms should be installed in circulation spaces (e.g. hallways and landings) and ionization chamber-based smoke alarms may be more appropriate if placed where a fast burning fire presents the greater danger to occupants (such as in living or dining rooms). Additionally, optical detectors are less affected by low levels of 'invisible' smoke that often cause false alarms.
8 Provided that the lifetime or duration of any standby power supply is not reduced, smoke alarms may be interconnected using radio-links.
9 The power supply options described here are all based on using the mains supply at the normal 240 volts. Other effective (though possibly more expensive) options exist, such as reducing the mains supply to extra low voltage in a control unit incorporating a standby trickle-charged battery, before distributing the power to the smoke alarms at that voltage.
10 Approved Document K, Protection from falling, collision and impact. DETR and the Welsh Office 1998.
11 BS 6399 Loading for buildings, Part 1 : 1984 Code of practice for dead and imposed loads.
12 A floor which is used as a bedroom but which is open on at least one side to some other part of the dwelling.
13 BS 5588 Fire precautions in the design, construction and use of buildings, Part 9 : 1989 Code of practice for ventilation and air conditioning ductwork.
14 BS 5588: Fire precautions in the design, construction and use of buildings, Part 4 : 1998 Code of practice for smoke control using pressure differentials.
15 Pipelines Safety Regulations 1996, SI 1996/825.
16 Gas Safety (Installation and Use) Regulations 1998 SI 1998/2451.

Chapter 7
Application to Buildings other than Dwellings

7.1 Introduction

Chapter 4 contains details of the general principles for the provision of means of warning and escape which apply to all building types. Additionally, in Chapter 5, a simple evacuation model is given for the design of means of escape in other residential buildings (hotels, hostels, halls of residence, small nursing homes and other similar institutional buildings where people sleep on the premises) and smaller, simpler buildings used for other non-residential purposes. This is based on the recommendations of sections 1 to 6 of AD B.

Because of its generality, the evacuation model outlined in Chapter 5 may prove to be too detailed for the design of means of escape in dwellings, whilst at the same time proving insufficient for the needs of more complex buildings.

The purpose of this chapter is to consider means of escape issues that are specific to particular building uses other than dwellinghouses, flats and maisonettes (dealt with in Chapter 6). Extensive use is made of alternative sources of information which are referred to, but are not actually contained within, Approved Document B1.

7.2 Houses in multiple occupation

7.2.1 Introduction

Under the Housing Act 1985 local authorities have the power to require means of escape in case of fire in houses which are occupied by persons not forming a single household (known as houses in multiple occupation or HMOs). In the case of HMOs with at least three storeys and a floor area (all storeys combined) of at least 500 m^2, LAs have a duty to ensure that means of escape are adequate. Before exercising their power or performing their duty in respect of means of escape from fire in an HMO, LAs must consult the fire authority.

Under section 352 of the Housing Act 1985 (as amended), a local housing authority may serve a notice requiring the execution of works to make an HMO fit for the number of occupants. Guidance on the standards expected is given in Department of the Environment Circular 12/92 (in England) or Welsh Office Circular 25/92 (in Wales). The guidance is not mandatory but LAs are expected to have regard to it when considering the exercise of their powers under section 352. A certain amount of flexibility is expected and LAs are reminded in the Circulars that they should try to attain achievable standards since excessively high standards may deter landlords from making accommodation available at all.

For HMOs with six or fewer residents the guidance in Chapter 6, section 6.2 to section 6.3.8 (which deals with means of escape from single family dwellinghouses) is adequate. An outline of the main provisions of Circular 12/92 is given below for HMOs which do not fall within this category.

7.2.2 *General considerations*

In determining specific measures appropriate to an HMO it is necessary to consider:

- The way in which individual occupancies and their escape routes are protected from the spread of products of combustion. This will involve an assessment of the need for fire-resisting construction of walls and floors, fire doors and sealing to prevent the spread of smoke and noxious gases.
- The maximum travel distances involved in escaping from each room to a final exit.
- The nature and suitability of the means of escape for the number and types of occupants (e.g. the steepness of stairways and the widths of doorways and corridors).
- The need for fire precautions such as fire warning systems, fire sensors and fire-fighting equipment.

7.2.3 *Means of escape provisions – a summary*

Circular 12/92 (C12/92) gives the following general guidance based on the number of storeys in the HMO:

(1) **2 storeys:** the stairway should be made a protected route. Generally, 30 minutes' fire protection is adequate except for the enclosure to an area of

higher fire risk. This is defined in C12/92 as a room or other area which, because of its function and/or contents, presents a greater risk of fire occurring and developing than elsewhere; such areas may include a large kitchen, boiler room, large storeroom and other similar risks. Such areas, and the enclosure to a stairway to a basement, should be provided with 60 minutes' fire protection. Automatic fire detection may be provided by single point smoke detectors (such as those recommended in AD B1 for dwellinghouses).

(2) **3 or 4 storeys:** the stairway should be made a protected route (see (1) above). An L2 fire warning system should be provided.

(3) **5 or 6 storeys:** three options are offered:
   (a) The stairway is a protected route (see (1) above) and an alternative means of escape is provided by way of an additional stairway which may be external to the building. An L2 fire warning system should be provided.
   (b) The stairway is a protected route (see (1) above), separated from the accommodation by protected lobbies, and a suitable upward escape route is available within the existing stairway. An L2 fire warning system should be provided.
   (c) The stairway is a protected route (see (1) above), separated from the accommodation by protected lobbies. An L2 fire warning system should be provided. Use for human habitation is restricted to five storeys.

(4) **Over 6 storeys:** most HMOs which come into this category are likely to be hostels (and/or student halls of residence) and are dealt with in section 7.3.

7.2.4 Management of HMOs

Regulation 10 of the Housing (Management of Houses in Multiple Occupation) Regulations 1990[1] imposes a duty on the manager of an HMO to maintain all means of escape from fire and all equipment provided by way of fire precautions, and to keep means of escape free from obstruction. The regulations also require the manager to cause signs indicating means of escape to be displayed.

7.3 Hostels, student halls of residence and buildings with similar uses

The buildings in this section are all categorised by the fact that the occupants are provided with a bedroom which is not equipped as self-contained accommodation. Therefore, kitchen/dining room, lounge and bathroom facilities are provided for the use of all occupants in common.

In most cases the guidance given in AD B1 (see Chapter 4) will be appropriate for hostel-type accommodation (purpose group 2(b)). Alternatively, more specific guidance may be found in DoE Circular 12/92 (sections 2.18 to 2.40). Generally, C12/92 follows the principles for means of escape described in Chapter 4; however, there are differences related to travel distances and these are summarised in Table 7.1. Additionally, the term 'area of higher fire risk' is used in C12/92 (see definition in section 7.2.3) and this has a significant influence on permissible travel distances.

Table 7.1 Travel distance limitations in hostels.

Escape from any point within:	Maximum travel distance for escape in one direction only (m)		Maximum travel distance for escape in more than one direction (m)	
	Within room	Total distance[a]	Within room	Total distance[a]
Sleeping area	9	18	18	35
Area of higher fire risk	6	12[d]	12	25[b]
Any other area	9	18	18[c]	35

[a] This is the distance from within the room to a point of access into a protected route, external route or final exit.
[b] Alternatively, up to 35 m when the total distance of travel is not wholly within the area of higher fire risk.
[c] Where there are at least three exits from the area and one of these is a final exit, this may be increased to 35 m.
[d] Alternatively, up to 18 m when the total distance of travel is not wholly within the area of higher fire risk.

7.4 Hotels and boarding houses

In most cases, the guidance given in AD B1 (see Chapter 4) will be appropriate when designing new hotel accommodation (Purpose Group 2(b)). Additionally, when altering or converting existing buildings to hotel uses, more specific guidance may be found in the *Guide to fire precautions in premises used as hotels and boarding houses which require a fire certificate*, published in 1991 by the Home Office Fire and Emergency Planning Department (the HOFD guide).

In general the HOFD guide follows the principles for means of escape which are discussed in Chapter 4; however, there are differences related to travel distances and these are summarised in Table 7.2. Additionally, the term 'area of high fire risk' is used in the HOFD guide but is not precisely defined. The best clue to the meaning of this term may be found in clause 12.5(d) of the HOFD guide where it asks for consideration of:

> 'certain areas which, due to their functions, may present a greater risk of fires occurring and developing than elsewhere, such as:

Table 7.2 Travel distance limitations in hotels and boarding houses.

Escape from any point within:	Maximum travel distance for escape in one direction only (m)		Maximum travel distance for escape in more than one direction (m)	
	Within room	Total distance[a]	Within room	Total distance[a]
Sleeping area	8	16	15	32
Area of high fire risk	6	12[b]	12	25[b]
Any other area	9	18	18[c]	35

[a] This is the distance from any point in the building to the nearest final exit, protected stairway, protected lobby or a door for means of escape in a compartment wall.
[b] It is permissible to use a combination of travel distances where the entire escape route is not wholly within an area of high fire risk (e.g. where the route starts in an area of high fire risk and then passes into an area of normal risk, the first part should not exceed 6 or 12 m as appropriate and the total route should not exceed 18 or 35 m as appropriate).
[c] Where there are at least three exits from the area and one of these is a final exit, this may be increased to 35 m.

- maintenance workshops,
- large kitchens,
- oil-filled boiler rooms,
- transformer or switchgear rooms, and
- basements used as sleeping accommodation, a dining room, bar area, or kitchen.'

7.5 Residential health care premises

7.5.1 Introduction

Residential health care premises such as hospitals, nursing homes and homes for the elderly differ from other premises in that they contain people who are bedridden or who have very restricted mobility. In such buildings, it is unrealistic to expect that the patients will be able to leave without assistance, or that total evacuation of the building is feasible.

Hence, the approach to the design of means of escape in these premises will demand a very different approach from that embodied in much of AD B1, and NHS Estates has prepared a set of guidance documents under the general title of *Firecode* for use in health care buildings. These documents are also applicable to non-NHS premises.

For the provision of means of escape in new hospitals the recommended guidance document to use is *Firecode. HTM 81 Fire precautions in new hospitals*[2]. For work that affects the means of escape in existing hospitals, see *Firecode. HTM 85. Fire precautions in existing hospitals*[3].

7.5.2 Means of escape in new hospitals

The following notes are based on the guidance in HTM 81 and deal only with the design of the means of escape. They are intended to be used at outline design stage in order to assess the feasibility of a possible hospital layout and should not be used for detailed design. For this, HTM 81 should be consulted.

The evacuation of an entire hospital in the event of fire would be an enormous exercise in which patients might be placed at risk due to trauma or their medical condition. Should evacuation become necessary it should be based on the concept of progressive horizontal evacuation, with only those people directly at risk from the effects of fire being moved.

Progressive horizontal evacuation

The principle of progressive horizontal evacuation requires that patients be moved from an area affected by fire through a fire-resisting barrier to an adjoining area on the same level. This area should be designed to protect the occupants from the immediate dangers of fire and smoke. The patients may remain there until the fire is dealt with, or await further evacuation to another similar adjoining area or down the nearest stairway. Should it become necessary to evacuate an entire storey, this procedure should give sufficient time for non-ambulant and partially ambulant patients to be evacuated down stairways to a place of safety. Therefore, every storey containing patient access areas within a hospital should be designed to provide for progressive horizontal evacuation.

Means of escape principles

The hospital should be designed and constructed so that patients, visitors and staff can move away from a fire to a place of temporary safety inside the hospital on the same level, from where further escape is possible to a place of safety outside the building.

HTM 81 provides guidance on means of escape by reference to:

- the potential for horizontal evacuation, which is achieved by dividing the storey into compartments and sub-compartments;
- the height above ground of the treatment area;
- travel distances and escape routes;
- the provision of an adequate number of stairways to facilitate vertical escape;
- emergency and escape lighting.

The principle of progressive horizontal evacuation will be achieved if the guidance below is followed.

Storeys up to 12 m above ground level with floor area less than 1000 m²

Every storey with a floor area of less than 1000 m^2 which contains patient access areas should:

- not contain more than 30 patients;
- be divided into at least two compartments;
- if providing sleeping accommodation, have no more than 20 beds in each compartment.

In storeys above ground floor level where sprinklers are installed, the fire resistance of the compartment walls may be reduced to 30 minutes (integrity and insulation).

Storeys up to 12 m above ground level with floor area more than 1000 m²

Every storey up to 12 m above ground, which has a floor area of more than 1000 m^2, and which contains patient access areas, should be divided into at least three compartments (one of these compartments may be a hospital street – see later section). In storeys above ground floor level where sprinklers are installed, the fire resistance of the compartment walls may be reduced to 30 minutes (integrity and insulation).

Storeys over 12 m above ground level

Every storey over 12 m above ground which contains patient access areas should:

- be divided into at least four compartments;
- where no hospital street is provided, each compartment should have a minimum floor area of 500 m^2;
- where one of the compartments is a hospital street, the area of the hospital street may be less than 500 m^2.

Where sprinklers are installed, the minimum floor area of each compartment may be reduced to 350 m^2.

Compartmentation

In a fire emergency each compartment should be large enough to accommodate the designed occupancy of the most highly occupied adjoining compartment, as well as its normal occupants.

The maximum size of compartment permitted by HTM 81, although appropriate for fire containment, is too large if the area contains patient access areas. In a fire emergency, a large number of patients could be overcome by fire, smoke and toxic gases. Therefore, in order to limit the number of patients who may be affected by a fire, compartments containing patient access areas should be divided into smaller sub-compartments.

A compartment should be sub-compartmented if:

- it has a floor area exceeding 750 m^2; *or*
- it contains departments to which more than 30 patients will have access at the same time; *or*
- it contains sleeping accommodation for more than 30 patients.

Sub-compartments should be enclosed by walls having a minimum of 30 minutes' fire resistance, and which terminate at the underside of either:

- a compartment floor; *or*
- a roof; *or*
- a non-demountable and imperforate ceiling which has a minimum of 30 minutes' fire resistance.

All openings in sub-compartment walls should be protected to provide a minimum of 30 minutes' fire resistance.

Exits from compartments

Exits from compartments should be direct to a circulation space; however, for ward bedrooms only, it is acceptable to escape directly to:

- an adjacent ward bedroom in an adjoining compartment or sub-compartment; *or*
- a circulation space in an adjoining compartment or sub-compartment.

Where a storey is divided into three or more compartments, there should be at least two alternative exits from each compartment, which provide horizontal escape to adjoining but separate compartments. Alternative exits should be located:

- as far apart as practical;
- if possible, in opposite walls;

so that in a fire situation at least one exit will always be available.

Escape routes may be located across flat roofs provided that:

- the roof and its structure forming the escape route are fire-protected;
- any walls within 3 m of the escape route contain no unprotected areas, are non-combustible and contain no extract ducts.

Travel distances

Where escape is possible in one direction only, the maximum travel distance before there is a choice of escape routes should not exceed 15 m.

Any part of an enclosed escape route where escape is possible in one direction only should be protected by 30 minute fire-resisting construction, and any glazing located in the enclosure should provide a minimum period of fire resistance of 30 minutes (integrity and insulation). Where a sprinkler system is installed the requirement for insulation may be omitted.

Escape from an inner room via an access room is permitted, provided the access room is not a fire hazard room (a room or other area which, because of its function and/or contents, presents a greater hazard of fire occurring and developing than elsewhere).

The maximum travel distance within a compartment should not exceed 60 m to:

- each of two adjoining compartments; *or*
- an adjoining compartment and to a stairway or final exit.

The maximum travel distance from any point within a sub-compartment should not exceed 30 m to:

- an adjoining compartment or sub-compartment;
- a stairway or final exit.

The 15 m single direction escape travel distance referred to above is included within the maximum travel distance and is not additional to it.

Hospital streets

Many hospitals incorporate the concept of the hospital street as the main communication route, although this is not an essential requirement. The hospital street provides an essential link between hospital departments and stairways and lifts and is the main circulation route for staff, patients and visitors. A hospital street may be considered to be a special type of compartment which connects final exits, stairway enclosures and department entrances. From a fire safety aspect, it has two main functions:

- the fire brigade may use it as a fire-fighting bridgehead; *and*
- if a fire within a department cannot be brought under control, it enables the occupants of the department to be evacuated via the hospital street to parts of the hospital not affected by the fire.

A hospital street should:

- be constructed to the same fire-resisting standards as a fire compartment;
- have a clear width of at least 3 m;
- be divided into at least three sub-compartments, with each not exceeding 30 m in length;
- at ground floor level, have at least two final exits located:
  - at all extremities of the hospital street;
  - so that the maximum travel distance between final exits measured along the length of the hospital street does not exceed 180 m;
  - so that the maximum distance from a compartment exit to a final exit does not exceed 90 m;
- at upper levels, have access to at least two stairways, each in separate sub-compartments, located so that:
  - the maximum distance between stairways is not greater than 60 m;
  - the maximum travel distance in one direction within the street does not exceed 15 m;
  - the maximum distance from a compartment exit to a stairway does not exceed 30 m;
- not contain any other accommodation apart from sanitary accommodation.

Entrances from the hospital street to adjoining compartments should:

- not be located in the same sub-division of the hospital street as entrances to stairways and lift enclosures;
- be positioned so that an alternative means of escape from each compartment is always possible.

Stairways should be located so that the maximum travel distance from the exit from the stairway enclosure to the final exit of the hospital street is no more than 60 m.

Widths of escape routes

In departments and wards, the width of the circulation spaces required for moving beds and patient trolleys will normally be adequate for escape purposes. In other parts of the hospital the widths of escape routes will be determined by the number of people who would normally be expected to use them in an emergency. In such areas, the minimum clear width of escape routes should be 1100 mm for up to 200 people. Where the number of users exceeds 200, an additional 275 mm of width should be allowed for each additional 50 people.

Vertical escape

In a fire emergency in a hospital, it is likely that any available stairway will be used. Therefore, all stairways should be designed as escape stairways. The stairways should be located in the hospital street, where this is provided. Where this is not the case, stairways should be provided to each compartment.

The number of stairways provided should be in accordance with the numbers shown in Table 7.3 and the stairways should be located so that an alternative means of escape is always available from every compartment and sub-compartment. External stairways should not be provided for escape purposes.

Table 7.3 Provision of stairways in hospitals.

Number of patient beds on any one upper storey	Number of stairways
1–100	2
101–200	3
201–300	4
301–400	5

In the event of fire at least one stairway must always be available for evacuation purposes. Therefore, stairways should always be remote from each other and this can be achieved either by distance or by separation using fire-resisting construction.

The guidance in HTM 81 is based on the assumption that the provision of lifts specifically for the evacuation of patients is not considered necessary. Therefore, all stairways to areas which provide sleeping accommodation should be designed to permit the evacuation of patients on mattresses (mattress evacuation). This has implications for the width of such stairways, which should be in accordance with the widths shown in Table 7.4. Various stairway widths and associated landing widths and depths are given in the table, all of which allow mattress evacuation but vary in their capacity to enable pedestrian passing when the stairway is being used for mattress evacuation. The dimensions refer to clear widths between handrails.

The width of stairways to areas which do not include patient sleeping

Table 7.4 Width (mm) of stairways and landings for mattress evacuation.

Stair width	Minimum landing width	Minimum landing depth
1300	2800	1850
1400	3000	1750
1500	3200	1550
1600	3400	1450
1700	3600	1400
1800	3800	1350

accommodation should be assessed as shown in the section 'Widths of escape routes' above.

All stairways should terminate at ground level and:

- provide access directly to the outside via a final exit; *or*
- discharge to a fire-protected route from the base of the stairway to the outside, which contains no accommodation, except that permitted for a protected shaft; *or*
- discharge to a hospital street.

Final exits

Although some final exits may be provided only for escape purposes, many of the final exits within hospitals will also be available for everyday use and, therefore, should be designed to permit access for people with restricted mobility. It should be recognised that often there are conflicting requirements for means of escape and security of the hospital, so any solution should be agreed in advance between the enforcing authorities, the hospital management and its security advisors.

Final exit doors:

- should open outwards and should never be provided with locks requiring a key for opening;
- from patient access areas should not be provided with a step, and should open onto a level area for a distance of 1 m;
- if automatic, should be freely openable by hand under any condition, including power failure; otherwise, adjacent non-automatic outward opening doors should be provided.

External escape routes

If the need arises in a fire emergency to evacuate an entire hospital or part of a hospital, adequate external assembly positions should be available. Suitable positions may be roadways, hard standings or suitably designed parts of the landscaping.

Although this is an operational requirement which cannot be enforced through the current Building Regulations, the following points should be taken into account when designing external escape routes:

- the location of assembly positions to permit access for ambulances;
- the provision of adequate artificial lighting;
- the provision of adequate paved footpaths and dropped kerbs to the assembly points;
- the gradients of external escape routes;
- the proximity of external escape routes to the external wall of the hospital.

7.5.3 *Means of escape in residential care premises*

Guidance on the provision of means of escape in existing residential care premises is provided in *Draft guide to fire precautions in existing residential care premises*[4]. This document is currently under review so will not be discussed in detail in this book. It covers means of escape in children's homes, community homes, homes for the elderly and homes for the mentally ill and mentally and physically handicapped (including homes that provide nursing care, but which have a greater affinity to residential homes than to hospitals).

Alternatively, guidance on the provision of means of escape in new and existing residential care premises, which is acceptable to all building control bodies, is contained in *Fire Safety in Residential Care Premises*[5]. The following notes are based on this document and deal only with the design of the means of escape. They are intended to be used at outline design stage in order to assess the feasibility of a possible layout and should not be used for detailed design. For this, the full document should be consulted.

Design principles

Fire Safety in Residential Care Premises (FSRCP) is limited to the design of residential care premises where resident or staff bedrooms are located up to three storeys above ground level or access level. The guidance is not appropriate for premises with bedrooms above this level.

The guide treats the means of escape as being in a series of stages:

- **Stage 1** – horizontal escape out of the room of origin of the fire.
  It is likely that most rooms will only have one exit; therefore, this will

normally be a single direction of escape. Stage 1 can also include any horizontal travel outside the room of origin in a dead-end situation.

- **Stage 2** – the part of the escape route from Stage 1 to a place of relative safety, such as a sub-compartment[6], protected stairway or external exit. This will involve passing through a fire-resisting barrier (e.g. a wall to a sub-compartment or a protected shaft enclosing a stairway) and there should always be the opportunity for an alternative direction of escape in this stage. In fact, a form of progressive horizontal evacuation may be incorporated into the design so that stage 2 can lead to a temporary refuge in a sub-compartment without a significant change in levels. If this strategy is adopted in the design there should be the potential for vertical escape to the ground floor should that become necessary.
- **Stage 3** – vertical escape down to the ground floor or access level and safe exit from the foot of the stairway to a safe place away from the building.

Travel distances

Travel distance is defined in FSRCP as:

> 'the distance to be travelled by a person from any point within the floor area to the nearest adjoining sub-compartment, stairway in a protected shaft, or external exit, having regard to the layout of walls, partitions, fittings and furniture.'

In general, the maximum travel distance for a single direction of escape in stage 1 should not exceed 10 m. This may be relaxed for two-storey buildings, where the first floor is a single sub-compartment and the stairway is not protected, provided that no point on the first floor is more than 20 m from the final exit from the building.

The maximum recommended travel distance from any point in the building to a place of safety or relative safety (i.e. the nearest adjoining sub-compartment, stairway in a protected shaft, or external exit) is 20 m. This can be made up of a single direction of escape of up to 10 m, at which point there should be alternative routes for the balance of the 20 m until a place of safety or relative safety is reached. Additionally, no bed situated on an upper floor should be more than 40 m from a protected stairway (see 'Stairways' section below).

Widths of escape routes

In new buildings, escape routes should have the following minimum clear widths between handrails:

- corridors – 1400 mm;
- stairways – 1200 mm.

Stairlifts which reduce the minimum width below 1200 mm are not permitted; however this guidance predates the changes to AD B1 which allow the guide rail of a stairlift to be ignored when assessing the width of the stairway (see section 5.2.4).

Stairways

In medium and large premises (i.e. those with more than three residents) there should be enough stairways provided within 40 m of any bed on an upper floor to ensure that every sub-compartment is provided with an alternative means of escape to the ground floor. All such stairways in new buildings should be located within protected shafts. In existing buildings, stairways may be placed externally, provided they are suitably protected from fire in the building. There are no requirements for premises with three or fewer residents. Where all points in a room on an upper floor are within 10 m of a protected stairway, there is no need to provide an alternative means of escape to the ground floor.

Where a building has only one stairway (whether external or internal) it should be constructed of non-combustible materials or materials of limited combustibility.

Final exits

All protected stairways should:

- have direct access to; *or*
- lead via a 30 minute fire protected route which is free of combustible material to

a place of safety located away from the building at ground level in the outside air. Where two or more protected stairways are provided in a building, the escape route from the foot of only one of the stairways may contain furniture.

7.5.4 *Unsupervised group homes for mentally impaired or mentally ill people*

If an existing house of not more than two storeys is converted for use as an unsupervised group home for not more than six mentally impaired or mentally ill people it may be regarded as a dwellinghouse (purpose group 1(c)) if the means of escape are provided in accordance with the guidance in *Firecode.*

HTM 88. Guide to fire precautions in NHS housing in the community for mentally handicapped (or mentally ill) people[7].

The guidance in HTM 88 may be summarised as:

- all stairways, including associated halls, landings and corridors should be enclosed with fire-resisting construction;
- any corridor not associated with a stairway should be enclosed with fire-resisting construction;
- stairways should not pass through nor be part of any room;
- stairways should give direct access to a final exit;
- single storey properties should have their entrance halls and associated corridors separated from other accommodation by fire-resisting construction;
- bedrooms constructed as inner rooms should have an alternative means of escape;
- corridors not directly connected to a final exit should have their own final exit.

New buildings in this category should be regarded as being in purpose group 2(b), Residential (other), (like hostels, hotels and boarding houses) and should follow the guidance in Chapter 4.

7.6 Small premises

The guidance given in Chapter 5 is well suited to the design of a range of residential and non-residential buildings. However, it may prove to be unduly restrictive in the design of small shops, offices, industrial, storage and other similar premises with its insistence on the provision of alternative means of escape and protected stairways.

The small size of such premises puts limits on the number of people who can use them at any one time. Therefore, it should be possible for the occupants to reach a single entrance/exit quickly in an emergency and the limited size of the premises ought to enable clear vision of all parts when undivided, thereby ensuring early warning. Thus, it may be possible to reduce the number of exits and stairs and, in certain cases, to omit protection to the stairway(s) altogether.

Such relaxations will not apply where the sale, storage or use of highly flammable materials is involved, as rapid evacuation of such premises is essential in the event of a fire. Additionally, in covered shopping complexes, the size of small units that may be served by a single exit is further restricted. This is dealt with in BS 5588 : Part10 (see section 7.10).

General recommendations

The following general recommendations apply in place of those given in Chapter 5:

- The premises should be in a single occupancy and should not comprise more than a basement, a ground floor and a first storey. No storey should have a floor area exceeding $280\,m^2$.
- Any kitchen or other open cooking arrangement should be sited at the extremity of any dead-end, remote from the exit(s).
- The planned seated accommodation or the assessed standing accommodation for small premises consisting of a bar or restaurant should not be greater than 30 persons per storey. This figure may be increased to 100 persons for the ground storey if it has an independent final exit.
- Floor areas should be generally undivided (except for kitchens, ancillary offices and stores) to ensure that exits are clearly visible from all parts of the floor areas.
- Store rooms should be enclosed with fire-resisting construction.
- Any partitioning separating a kitchen or ancillary office from the open floor area should be provided with sufficient clear glazed areas, to enable any person within the kitchen or office to obtain early visual warning of an outbreak of fire. Alternatively, an automatic fire detection and alarm system may be provided in the outer room.
- Travel distances should not exceed the values given in Table 7.5.
- Where two or more exits or stairs are provided, they should be sited so that they permit effective alternative directions of travel from any relevant point in a storey.

Provision of stairs

As a general rule, two protected stairways should be provided to serve each storey. However, the following small premises may be served by a single protected stairway provided that the conditions listed below are met:

Table 7.5 Maximum travel distances in small offices (and certain other small premises).

Storey	Maximum travel distance[a] to nearest storey exit (m)	Maximum direct distance[b] to nearest storey exit (m)
Ground storey with single exit	27	18
Basement of first storey with single stairway	18	12
Storey with more than one exit/stairway	45	30

[a] The distance of travel in small premises with an open stairway is measured to the foot of the stair in a basement or to the head of the stair in a first storey.
[b] For design purposes only, direct distances may be used.

- any small premises except bars and restaurants;
- an office building comprising not more than five storeys above the ground storey, provided that:
  - the travel distance from every point in each storey does not exceed 18 m (or 12 m if direct distance); *and*
  - every storey at a height greater than 11 m has an alternative means of escape.
- a factory consisting of not more than two storeys above the ground storey (for low risk buildings) or one storey above the ground storey (for normal risk buildings), provided that the travel distance from every point in each storey does not exceed 18 m (or 12 m if direct distance);
- process plant buildings with an occupant capacity not exceeding 10.

In certain circumstances a stairway in small premises may be open (i.e. unprotected) if it does not connect more than two storeys and delivers into the ground storey not more than 3 m from the final exit. For such an arrangement to be permitted it will be necessary for the storey served by the open stair either:

(1) to be served also by a protected stairway; *or*
(2) if it is a single stair
   - for the floor area in any storey not to exceed 90 m^2; *and*
   - if the premises contains three storeys (i.e. basement, ground and first floors) for the stair serving either the first or basement storey to be enclosed with fire-resisting construction at the ground storey level and to discharge to a final exit without the need to re-enter the ground storey (see Fig. 7.1).

7.7 Offices and other buildings with exits in a central core

The guidance given in Chapter 5 is suitable for offices and other buildings where the stairways are situated adjacent to external walls. This can result in the building being relatively narrow on plan which may not allow efficient use to be made of certain development sites where there is a call for a deeper, squarer style of design.

It is possible to site exits to stairways around the central core of a building, thereby allowing more efficient use of space, provided that certain conditions are met. Therefore, buildings with more than one exit in a central core should be planned so that:

- the storey exits are remote from one another; *and*
- no two exits are approached from the same lift hall, common lobby or undivided corridor, or linked by any of these.

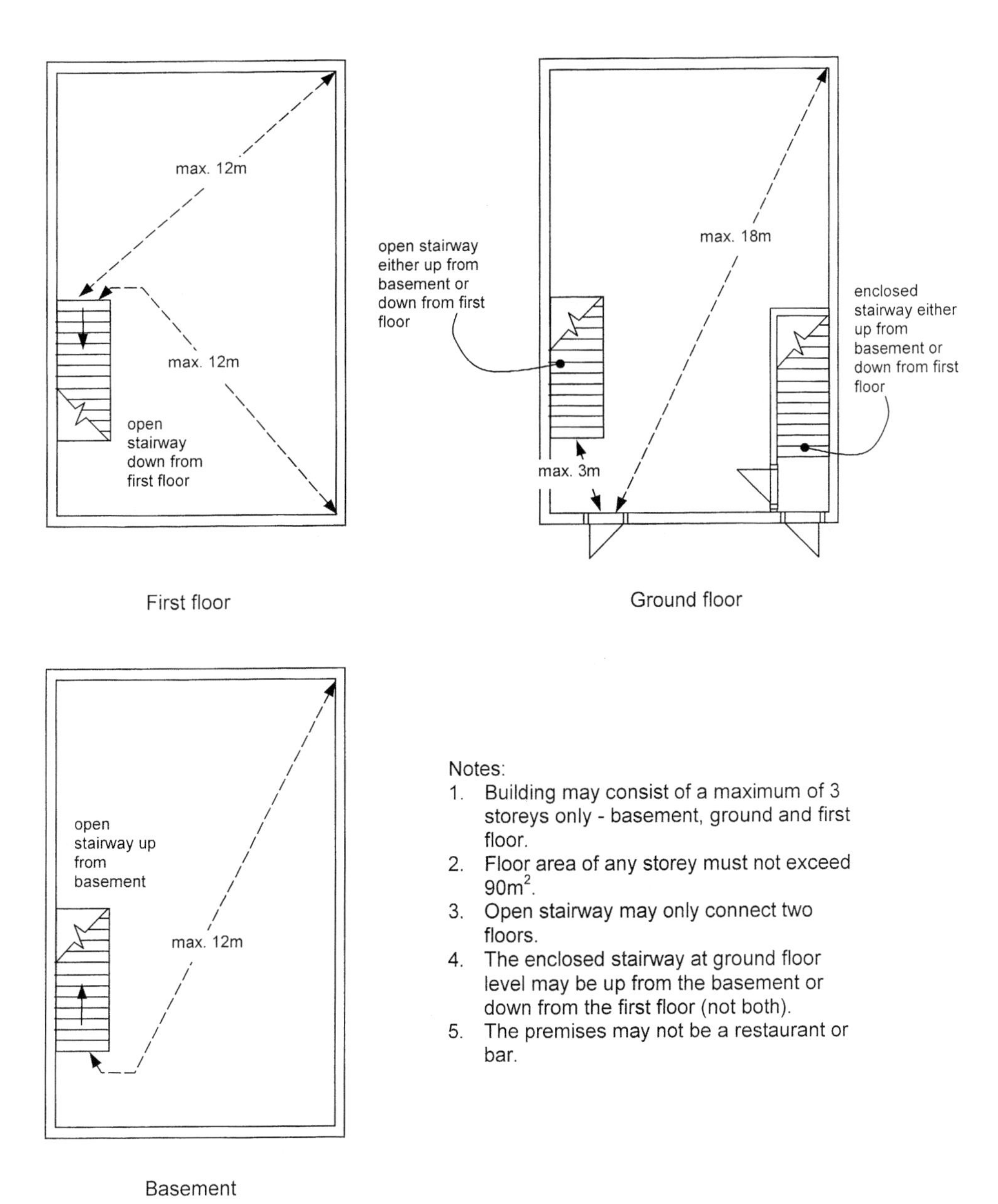

Fig. 7.1 Small premises – maximum travel distances and protection to stairs.

Figure 7.2 illustrates a possible arrangement for a central core which satisfies the above conditions and although this is most likely to be used in the design of office blocks, the principles apply equally to buildings in other purpose groups.

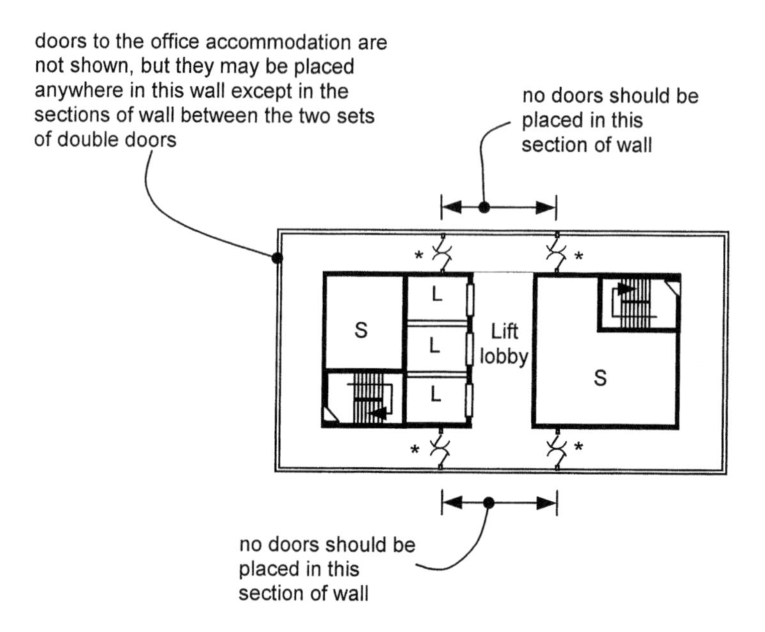

Key
* FD20 self-closing fire door (20 minutes integrity)
L Lift
S services, toilets etc.

Fig. 7.2 Corridor layout: building with central core.

7.8 Schools and other educational buildings

As a result of the coming into force of the Education (Schools and Further and Higher Education)(Amendment) Regulations 2001[8] on 1 April 2001, schools in England are no longer exempt from the Building Regulations. This eventuality had been foreseen for a number of years and had resulted in amendments to Approved Document B (especially B1) as long ago as the autumn of 1999.

Formerly, local authority schools had been exempt from the provisions of the Building Regulations and means of escape design had been subject to the guidance contained in the DES publication *Building Bulletin 7. Fire and the design of educational buildings*.

Much of the guidance contained in Chapter 5 is relevant to schools and other educational buildings; however, a number of points are highlighted below, which are of particular reference to the design of means of escape in such premises.

Limitations on travel distance

Educational buildings fall within purpose group 5, Assembly and recreation, and the maximum recommended travel distances are as follows:

- where escape is possible in one direction only – 18 m;
- where escape is possible in more than one direction – 45 m.

Widths of escape routes and exits

In schools, the minimum width of corridors in pupil areas should be 1050 mm, where escape is possible in more than one direction. In dead-ends, the minimum width should be increased to 1600 mm.

Single escape stairs

Where a school building:

- is not excluded from having a single escape route by virtue of steps 3 and 4 in Chapter 5; *and*
- has travel distances on every storey which are within the limits for means of escape in one direction,

it may be served by a single escape stair if it has no floor more than 11 m above ground, provided that any storeys above the first floor level are only occupied by adults.

Additionally, in a two-storey school building (or part of a building) served by a single escape stair, the first floor should not contain more than 120 pupils plus supervisors. Classrooms and stores should not open directly onto the stairway and the first floor should not contain a place of special fire hazard (laboratory, technology room with open heat sources, kitchen or store for PE mats or chemicals).

7.9 Assembly and recreation buildings

A number of problems are associated with the design of means of escape in assembly buildings, such as sports grounds, theatres, concert halls and conference centres, which are unlikely to occur in other building uses:

- large numbers of people are likely to be present;
- many of the occupants will be unfamiliar with their surroundings;
- there will be a wide range in terms of age and mobility; *and*
- the presence of fixed seating may limit the ability of people to escape.

The general guidance given in Chapters 4 and 5 is relevant in the design of means of escape from assembly buildings; however, as an alternative to this, AD B1 suggests the use of BS 5588 : Part 6[9]. This standard covers all aspects of fire precautions in the design, construction and use of assembly buildings, but the following notes deal only with means of escape.

General considerations

In assembly buildings, escape routes should be designed so that it is possible to evacuate the whole building in the event of fire. Escape routes will usually be made up of two principal parts:

- the travel distance within an enclosed space; *and*
- where appropriate, the travel distance within a protected space leading to a final exit.

Furthermore, if seating is provided in rows or blocks, or if exhibition stands are provided, the gangways will also have a bearing on the location of exits.

Therefore, where possible, escape routes from public areas should give direct access to a final exit. Where design considerations do not allow this, escape routes may pass through a fire-protected space or a protected corridor and protected space before reaching a final exit. Ideally, uniformly distributed alternative exits should always be provided from every public area and these should be planned so that they do not pass through the same protected space or corridor.

In theory, ancillary accommodation should not form part of any escape route designed for the public. In practice, there may be situations where escape routes for the public incorporate some types of ancillary accommodation, such as bars, souvenir shops or fast food outlets. This can be dealt with in a number of ways:

- separate the ancillary accommodation from the escape route by fire-resisting construction; *or*
- incorporate other fire safety measures into the design (e.g. sprinklers and fire shutters and smoke control operated by smoke detectors) to ensure the viability of the escape route; *or*
- provide a sufficient number of alternative escape routes from the public area

so that the escape route containing the ancillary accommodation can be discounted and the remaining routes can accommodate the full capacity.

High fire risk ancillary accommodation (such as kitchens) must be separated from the escape routes by fire-resisting construction.

Number of escape routes

In assembly buildings there will usually be a heavy concentration of people. Therefore, it is important to relate the number of escape routes to the number of people involved and to their distribution, so that congestion can be avoided and wide dispersion can be achieved. Some assembly areas may be multi-purpose in that they can be used with both open floor areas and seating in rows. If so, they must be designed to cater for both eventualities (usually by meeting the travel distances for both configurations).

The recommendations for numbers of escape routes contained in Chapter 5 are identical to the recommendations of BS 5588 : Part 6.

Travel distances

The recommended travel distances contained in Chapter 5 apply equally if BS 5588 : Part 6 is used as the basis of the design. However, a number of additional considerations which are specifically related to assembly buildings need also to be considered.

- Where a stage is provided, the direction of travel should be away from the stage (apart from any initial travel along a seatway).
- For multi-purpose halls, when these are used for exhibitions, the erection of display stands will create greater travel distances than if the hall is used where the floor plan is open. In such cases direct distances (i.e. two-thirds of the travel distance) should be used as the basis of the design.
- There may be cases in arenas and large exhibition halls where the quoted travel distances will be unrealistically restrictive. In such cases either of the following alternatives may be adopted:

  (1) provide at least two fire-protected subways leading from the central areas of arenas and exhibition halls to final exits. Access to the subways should be carefully located to eliminate or reduce excess travel distances and to avoid all courts and pitches in sports arenas. For example, the subways could be located at the 'D' ends of tracks in athletics arenas; *or*

  (2) provide a smoke control system which complies with the provisions detailed in clause 30.2.3 of BS 5588 : Part 6.

Widths of escape routes

The rules for calculating exit widths contained in Chapter 5 are also applicable when using BS 5588 : Part 6. This also applies to the rules concerning discounting, whereby it is assumed that one exit will be made impassable due to fire and the remainder of the exits will need to cope with the occupants of the area being evacuated. Discounting is not necessary if there are three or more exits provided from the assembly area and they are all final exits, since the fire cannot affect any part of the escape routes outside the space and any fire should be visible to the occupants at an early stage.

Occupancy numbers

Where the assembly building is likely to be used without fixed seating (for exhibitions, pop concerts, etc.) it will be necessary to calculate the maximum number of occupants for means of escape purposes. As explained in Chapter 5, this will have a bearing on the number of exits and escape routes which will need to be provided, and their respective widths. Table 7.6 contains floor space factors which are applicable to assembly buildings and are in addition to those contained in Table 5.1. When using Table 7.6 the net internal area of the space should be used (i.e. the total floor area excluding any space occupied by stair enclosures, lift wells, escalators, accommodation stairs and toilet accommodation).

Table 7.6 Additional floor space factors for assembly buildings.

Floor space factor (m^2 per person)	Type of accommodation/use of room	Notes
0.4 to 0.5	Individual seating	
0.3	Bench seating	If the number and length of benches is known, a factor of 450 mm per person should be used.
1.1 to 1.5	Restaurants and similar table and chair arrangements around a dance area	
1.2	Ice rinks	
1.5	Exhibition	Alternatively, a factor of 0.4 m^2 may be used over the gross area of gangways and other clear circulation space between stalls and stands.
9.5	Bowling alley	

Discharge from stairs and final exits

The rules for dealing with the discharge from stairs and final exits contained in Chapter 5 (see step 9) are also applicable when using BS 5588 : Part 6. One important variation in respect of assembly buildings lies in the fact that it is common practice to provide stairs from the various assembly levels which discharge into the main entrance foyer. Foyers are areas which will be familiar to the public and they are likely to use them in an emergency, therefore the following additional considerations may be necessary in their design:

- The foyer should be enclosed with fire-resisting construction if it receives escape routes from other tiers in the building.
- Where a number of different auditoria discharge into one foyer (such as from the different cinemas in a multiplex development) the foyer should be enclosed with fire-resisting construction, and protected lobbies should be provided between the foyer and the escape routes which discharge into it.
- There should be other escape routes provided from each of the areas which discharge into the foyer and they should lead directly to a place of safety.

Seating

It is usual practice to limit the number of seats in a row since a good seating layout can assist the orderly movement to exits. Where gangways are provided at each end of the rows of seating the numbers are relatively unimportant if travel distances are adhered to and generous seatways are provided. Table 7.7

Table 7.7 Seatway widths and allowable numbers of seats in a row.

Seatway width (mm)	Maximum number of seats in a row with gangway on:		Notes
	One side	Both sides	
300–324	7	14	1. Seatway widths should not be less than 300 mm and should be constant throughout the row.
325–349	8	16	
350–374	9	18	2. For automatic tip-up seats measure the seatway width from the back of one seat to the forwardmost projection of the seat behind.
375–399	10	20	
400–424	11	22	
425–449	12	24	3. Seatways in front of blocks of seating can be up to 900 mm wide without being treated as gangways.
450–474	12	26	
475–499	12	28	4. The slope of tiers of seating should not exceed 35° to the horizontal.
500 and over	12	Limited by the travel distance	

can be used to design seatways by relating the number of seats in a row to the seatway width.

Gangways

The design of gangways is a complex matter and will depend greatly on the use of the assembly building. BS 5588 : Part 6 contains a large number of recommendations which are related to particular design arrangements and may go beyond simple means of escape considerations. The following general comments can be made but for detailed advice, reference should be made to BS 5588 : Part 6, section 8.3.

- Gangways should be at least 1100 mm wide unless used by 50 people or less, when they can be reduced to 900 mm.
- Handrails are permitted to encroach upon the clear width, provided that they do not project more than 100 mm.
- The ends of all rows of seats should be aligned to maintain a uniform width of gangway.
- Storey exits within the body of the seating layout should be approached from the side by transverse gangways.
- In stepped tiers the step height should be between 100 mm and 190 mm and the number of uninterrupted steps in a gangway pitched at greater than 25° to the horizontal should not exceed 40.
- In stepped tiers there should not be a change of level between the seatway and the nearest step.

7.10 Shopping complexes

The guidance given in Chapter 5 is well suited to the design of means of escape within individual shop units and is also suitable for the design of escape stairs and final exits in shopping complexes. However, where shopping units open into a covered complex, the recommendations referred to above need to be varied because the units generally need to open directly onto the malls without any intervening compartmentation, in order for the complex to function satisfactorily. Shop fronts which are provided are unlikely to offer any degree of fire-resisting separation, and life safety is also affected by the large number of persons likely to be present, especially during peak trading hours and peak shopping seasons. For these reasons, reference should be made to BS 5588 : Part 10[10] in order to obtain specific guidance for the design of means of escape in shopping complexes. This standard covers all aspects of fire precautions in the design, construction and use of shopping complexes, but the following notes deal only with means of escape.

General considerations

Shopping malls are, in effect, a substitute for the open street during initial escape from a fire-affected shop unit. Additionally, because the main retail areas and the malls are often a single fire compartment, it is important to consider:

- the impact that a fire in one unit might have on other units within the same fire compartment; *and*
- the consequences of the fire for persons escaping from these other units.

These considerations result in the need for a package of interrelated fire safety features which will affect the design of the individual units as well as the design of the common areas.

In a fire or other emergency, shoppers will tend to leave by the route they know best, i.e. the shopping malls by which they entered the complex. This means that the entrances onto malls from shops and other units are effectively 'storey exits' for evacuation purposes, resulting in the need to keep the malls reasonably clear of smoke in the event of fire. This problem of smoke control can be dealt with in two ways:

(1) Extract hot smoky gases from the unit of fire origin. This may be impractical where there are many small units, as each unit would need to have an individual smoke control system.
(2) Allow smoke to enter the mall whilst it is at the hot and buoyant stage, then restrict its sideways spread by the use of curtains at ceiling level. These create smoke reservoirs from which smoke is extracted by natural or powered means at a rate which enables people to escape under the smoke layer with a reasonable degree of safety.

In the event of evacuation from a fire in a covered shopping complex the primary route of escape is via the mall, and the various fire safety features are intended to enable the mall to be considered as a street for escape purposes. Additionally, except in very small units, a secondary means of egress from each shop is generally needed, since the location of the fire might prevent people from escaping into the mall from the unit of fire origin and from the one opposite. Service corridors at the rear of the units may often prove suitable for use as a secondary route subject to adequate fire separation from other parts of the complex. When this is done, it is essential to install adequate management arrangements to ensure that the corridors are kept free from obstruction.

Basic design criteria

The basic criteria for determining the design of the means of escape serving units comprising a covered shopping complex are:

- Where possible provide an alternative means of escape (this does not apply to small units and kiosks which do not exceed 25 m^2 in total area and 5 m in depth).
- Where direct escape to a place of safety is not possible:
  - the travel distance in an unprotected escape route must be limited and lead to a protected escape route; *and*
  - the protected escape route needs to lead to a place of safety.
- The means of escape in individual shop units should be designed in accordance with the recommendations contained in Chapter 5 or in BS 5588 : Part 11. It should be noted that the exception which permits small shops to have a single exit (see section 7.6) does not apply if they are within a shopping complex. In this case they should be provided with alternative means of escape if they exceed 25 m^2 in total area and 5 m in depth.
- At least one alternative means of escape from any unit (other than a small unit or kiosk) should deliver either:
  - at a different level from that at which the entrance to the unit is situated; *or*
  - at the same level, but leading to a different final exit(s).
- Any exit from a unit that has an occupant capacity exceeding 300 persons should not discharge into a service corridor.

Travel distances in malls

Escape routes from the mall should be sited so that a person confronted by fire on entering the mall from a shop unit can make a safe escape through an alternative mall exit. For escape purposes, it is not considered acceptable for units to be used as an alternative escape route from a mall. Table 7.8 gives recommended travel distances for both covered and uncovered malls.

Widths of malls

The minimum width of a mall depends upon the need to avoid fire spread by radiation and the encroachment of flames across the mall, and the need to provide sufficient room for people to use the mall to escape to a mall exit. Consequently:

(1) To provide adequate separation against fire spread, a mall should be:
- not less than 6 m wide measured between the leasehold boundary or fascia of the units (whichever is the lesser) if covered; *or*
- not less than 5 m wide if uncovered; *and*

(2) To provide adequate capacity for escape:
- a covered mall with units on both sides should have an effective width

Table 7.8 Travel distances in malls.

Position of mall in complex	Travel in one direction (m)		Travel in more than one direction (m)		Notes
	Uncovered malls (1)	Covered malls (2)	Uncovered malls (3)	Covered malls (4)	
Mall at ground level	25	9	No limit	45	If travel is initially in one direction only, then the distance to the point at which travel is possible in more than one direction should not exceed the appropriate limit given in columns (1) and (2).
Mall not at ground level	9	9	45	45	The total travel distance to the nearest mall exit or storey exit (including the initial travel in one direction only) should not exceed the appropriate limit given in columns (3) and (4).

(i.e. that part which is unobstructed by planters, seating, kiosks and void openings, etc.) of not less than 6 m; *and*

- a mall with a void on one side (e.g. a galleried upper mall overlooking the level(s) below), or a mall with units on one side and an imperforate fire-resisting wall on the other, should have an effective width of at least 3 m.

(3) The aggregate width of any doors placed across a mall should not be substantially less than the width of the mall itself.

(4) Where it can be shown that the number of people using a mall for escape at a particular point can be safely accommodated, the effective widths for galleried upper malls may be reduced, provided that the unobstructed width is not less than 1.8 m. A similar approach should be adopted where the width of mall or galleried upper mall is reduced by an obstruction.

Widths of mall exits

When added together, the widths of all the mall exits from a particular section of mall (i.e. the aggregate mall exit width) must be adequate for total occupant capacity of that section. The width of each exit can then be calculated simply by dividing the aggregate mall exit width by the number of exits, although this must never result in an exit which is less than 1.8 m wide. The aggregate mall exit width may be calculated using the following equation:

$$W = 5(xy/z)\text{mm}$$

where
W is the required aggregate mall exit width (mm);
x is the width of mall section (m);
y is the length of mall section (m);
5 is the exit width per person (mm);
z (m^2) is the appropriate floor space factor (see below);

Generally, the floor space factor (z) will be 0.75; however, there are exceptions to this rule:

- where an area of the mall section has fixed tables and associated seating, a floor space factor of 1.0 should be used for that area;
- where the width of the mall section exceeds 8 m, a floor space factor of 2.0 should be used for the part of the mall section in excess of 8 m;
- where the shop unit exits served by the mall section would permit a greater population to enter the mall section than that calculated using the appropriate floor space factors, the mall exits should be designed to take the greater population.

For calculation purposes, display areas and mall furniture (other than fixed seating for food courts) should not be deducted from the area of the mall although permanent display features (such as fountains) may be deducted.

Example 1: if a section of mall is 6 m wide and 90 m long with an exit at each end, then the combined width of the two exits should be not less than:

$$(6 \times 90)/0.75 \times 5 = 3600\,\text{mm}$$

The minimum width of each mall exit is therefore 1.8 m (which is in accordance with the rules above).
Example 2: if a section of mall is 12 m wide and 90 m long with an exit at each end, then the combined width of the two exits should be not less than:

$$\{(8 \times 90)/0.75\} \times 5 = 4800\,\text{mm } \textit{plus } \{(4 \times 90)/2.00\} \times 5 = 900\,\text{mm}$$

Therefore the total width required = 5700 mm
The minimum width of each mall exit is therefore not less than 2.85 m.

7.11 Means of escape and atria

An atrium is defined in Approved Document B and in BS 5588 : Part 7[11] as 'a space within a building, not necessarily vertically aligned, passing through one or more structural floors'.

An atrium provides a route by which smoke and fire may spread throughout a building much more rapidly than in the equivalent non-atrium building. Additionally, the volume of smoke may increase greatly due to the entrainment of air into the rising plume. During this period, the large quantities of smoke and corrosive fumes which are produced may cause damage which is disproportionate to the scale of the initial fire, if allowed to spread throughout the building via the atrium.

Such a spread of fire and smoke can have a significant effect on the number of people initially at risk, the time available for escape and the activities of fire-fighters.

In order to obtain specific guidance on all aspects of fire precautions in the design, construction and use of buildings incorporating atria, reference should be made to BS 5588 : Part 7. The recommendations of this code provide a range of options related to the following:

- effective planning and protection of escape routes from areas threatened by fire;
- limitation of fire development by the provision of automatic suppression systems or the control of materials;
- provision of fire warning systems and, where appropriate, automatic fire detection systems;
- separation of the atrium from associated floor areas;
- provision of smoke, pressure and temperature control systems to maintain the effectiveness of escape routes and access for fire-fighters;
- effective management control.

The code does not preclude the incorporation of atria into any form of building design, provided that it can be explicitly demonstrated that the life safety objectives are achieved by adequate fire protection.

The following notes are based on the guidance in BS 5588 : Part 7, but deal only with means of escape.

Escape routes

In a building where the accommodation is separated from the atrium by smoke retarding construction, the means of escape from the accommodation should follow the guidance given in this chapter and in Chapter 5 for the equivalent non-atrium building.

In a building where the accommodation is open to the atrium (or not enclosed by smoke retardant construction), the following guidance is applicable:

- escape routes should not approach the atrium edge, therefore storey exits should be sited away from the atrium;
- maximum travel distances from the atrium to the nearest storey exit, and capacities of exits and of the means of escape generally, should all be in accordance with the recommendations of the appropriate guidance for the equivalent non-atrium building;
- in non-residential buildings (i.e. where the occupants are unlikely to be asleep) escape should be away from the atrium void, and the subsequent escape route should not pass within 4.5 m of the atrium void.

Balcony escape

Sometimes the means of escape will be via a balcony within the atrium, with no alternative route from the accommodation being available. This is possible if the following recommendations are applied:

- the building should be fitted throughout with sprinklers (unless these are not required for the smoke control design);
- at least two directions of escape should be available within the atrium with the travel distance to the nearest storey exit not exceeding 18 m;
- where the balcony is enclosed by smoke retarding but not fire-resisting construction, a temperature control system (e.g. a system using the principles of smoke exhaust ventilation, or sprinklers designed to reduce smoke temperatures) should be provided;
- where the balcony is open, any smoke layer should be confined to a level not less than 3 m above the highest balcony or bridge by means of a smoke exhaust ventilation system;
- where the balcony is enclosed, fire-resisting and smoke retarding construction should be to the same specifications as that provided for the atria;
- the fire load at the base of the atrium should be limited either:
  - by means of management controls on the quantities of combustible material that are present; or
  - by an effective automatic suppression system.

7.12 Disabled people

The methods of providing means of escape contained in the preceding sections of this book assume that building users are able-bodied and that the role of management in a fire is to ensure that the fire brigade is called and that any evacuation of the building proceeds to plan.

The assumption that people are independently capable of using steps and stairs for escape purposes is clearly inadequate when considering the safety of some disabled people. For them, evacuation may involve the use of refuges on escape routes and either assistance down (or up) stairways or the use of suitable lifts. Therefore, BS 5588 : Part 8[12] details the measures which are appropriate for the safety of disabled people from the moment they, and/or the building management, become aware of potential danger until a place of safety is reached.

Comprehensive management procedures can greatly assist the successful emergency evacuation of a building. Management procedures for disabled people need to include not only arrangements for assisting people in wheelchairs but also those with walking difficulties or certain other disabilities (such as impaired hearing or sight).

Refuges

In general, the access provisions of Approved Document M[13] should ensure that normal horizontal circulation/escape routes are accessible and do not present obstacles, such as isolated steps or narrow corridors, which might impede the progress of a disabled person in the event of a fire emergency. Additionally, the limitations on horizontal travel distances in the various fire codes discussed above mean that most disabled people should be able to reach a protected escape route or final exit without assistance.

However, disabled people who are reliant on a wheelchair for mobility will not be able to use stairways without assistance. Therefore, it may be necessary to provide refuges (i.e. temporarily safe spaces for disabled people to await assistance for their evacuation) to each protected stairway on all storeys except those:

- in small premises (see section 7.6) of limited height (where the travel distance to a final exit is so limited that the provision of refuges is unnecessary);
- where level access is provided directly to a final exit; *or*
- which consist exclusively of plant rooms.

A refuge needs to be large enough to accommodate a wheelchair user and allow ease of manoeuvrability. This means that the minimum space provided for a wheelchair in a refuge should be 900 mm × 1400 mm. A satisfactory refuge could consist of:

- an enclosure such as a compartment, or a protected lobby, corridor or stairway;
- an area in the open air, which is sufficiently protected (or remote) from any fire risk and is provided with its own means of escape, such as a flat roof, balcony, podium or similar place.

Refuges (and the wheelchair spaces within them) should be located so as not to adversely affect the means of escape provided in the building. This means that the wheelchair space should not reduce the width of the escape route. Where the wheelchair space is within a protected stairway, access to it should not obstruct the flow of persons escaping.

Refuges should be clearly identified by appropriate fire safety signs. Where a refuge is in a lobby or stairway, it is essential that the sign is accompanied by a blue mandatory sign worded 'Refuge keep clear'.

Examples of typical refuges in a building with and without an evacuation lift are shown in Figs 7.3 and 7.4.

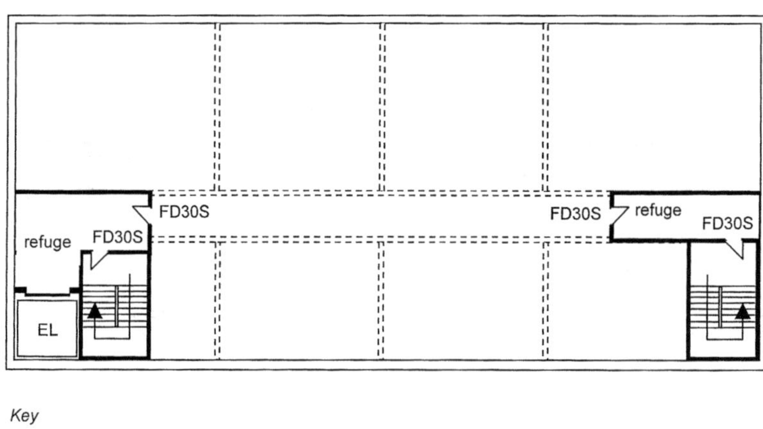

Key

FD30S self-closing fire door (30 minutes integrity) plus smoke seals

▬ 30 minutes fire-resisting construction

EL evacuation lift

Fig. 7.3 Refuges in buildings containing evacuation lift.

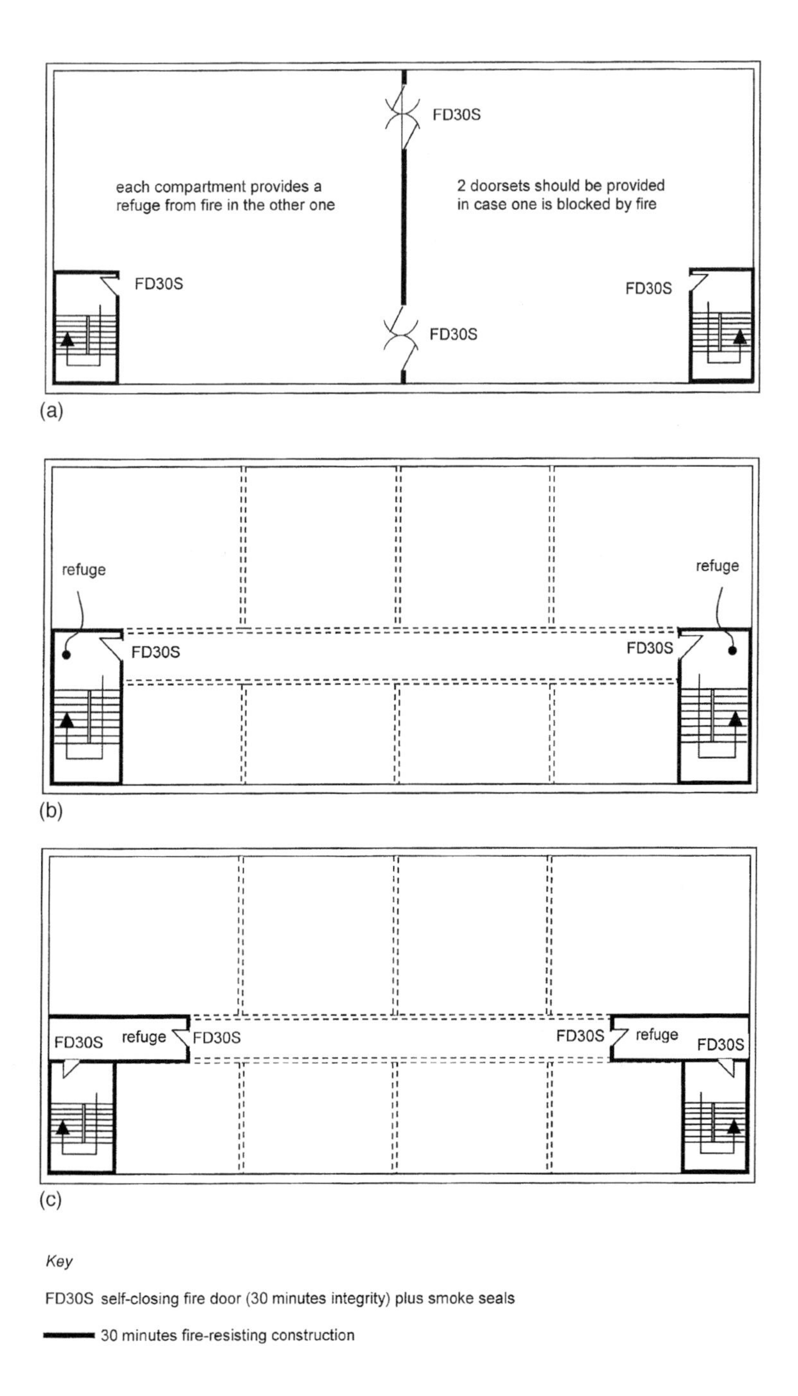

Fig. 7.4 Refuges in buildings not containing evacuation lifts. (a) Refuges within accommodation (no refuges within protected stairways); (b) Refuges within protected stairways; (c) Refuges with protected lobbies.

Stairways

In general, stairways which are designed in accordance with Approved Documents K and M will be adequate for use by disabled people in a fire emergency. Where installed, wheelchair stairlifts are not suitable for use as a means of escape. Wheelchair stairlifts that are provided for access should only be located within escape stairways if the effective width of the stairway is equal to or exceeds the width required for escape. The effective width will take into account the permanent incursion into the width of the flight of any part of the stairlift installation; however this guidance predates the changes to AD B1 which allow the guide rail of a stairlift to be ignored when assessing the width of the stairway (see section 5.2.4).

Evacuation lifts

Lifts are not normally used for means of escape since there is always the danger that they may become immobilised due to power failure and may trap the occupants. Although not absolutely essential, an evacuation lift may be provided as part of a management plan for assisting disabled people to escape, if the lift installation is appropriately sited and protected. It should also contain sufficient safety devices to ensure that it remains usable during a fire.

Fire-fighting lifts may be used for the evacuation of disabled people prior to the arrival of the fire service. Where such an arrangement exists, liaison with the fire authority to co-ordinate procedures for the use of fire-fighting lifts is essential.

An evacuation lift should:

- conform to the relevant recommendations of BS 5810[14] and BS 5655[15]: Part 1 or Part 2;
- be clearly identified with appropriate fire safety signs;
- be situated within a protected enclosure consisting of the lift well itself and a protected lobby at each storey served by the lift;
- be provided with a protected route from the evacuation lift lobby at the final exit level to a final exit;
- be provided with a switch clearly marked 'Evacuation Lift' situated next to the lift landing door at the final exit level. Unauthorised operation of the switch may be prevented by the use of a key operated switch or by placing the switch in a glass-fronted box. Operation of this switch should:
  - cause the evacuation lift to return to the final exit level
  - isolate the lift landing call controls
  - cause the lift to operate only in response to the lift car control panel;
- include a communication system designed to enable management to identify the location of disabled people needing evacuation;

- be powered by a primary electrical supply obtained from a sub-main circuit exclusive to the lift and independent of any other main or sub-main circuit. Other lifts in the same well may be fed from the same primary supply, provided that a fault occurring in any of the other lifts (or their power supplies) does not affect the operation of the evacuation lift;
- be provided with an alternative power supply (using fire-protected cables and equipment) such as an automatically started generator, a privately owned distribution system that would not be affected by a fire in the building or a separately fused protected circuit fed directly from the main incoming electrical supply to the building located in a fire protected enclosure. The requirement for an alternative power supply does not apply to hydraulic lifts serving two storeys only, the lower of which contains a final exit.

Management procedures in case of fire

The successful emergency evacuation of a building can be greatly assisted by comprehensive management procedures. This applies whether the occupants of a building are disabled or not, but has particular significance when evacuation lifts are installed as part of the emergency evacuation plan. BS 5588 : Part 8 should be consulted for advice on the development of suitable management procedures, including the essential requirement for independent communication between the occupants and evacuation management personnel.

7.13 References

1. Housing (Management of Houses in Multiple Occupation) Regulations 1990 (SI 1990/830).
2. *Firecode. HTM 81. Fire precautions in new hospitals.* (NHS Estates) HMSO 1996.
3. *Firecode. HTM 85. Fire precautions in existing hospitals.* (NHS Estates) HMSO 1994.
4. *Draft guide to fire precautions in existing residential care premises.* (Home Office Fire Department) HMSO 1983.
5. *Fire Safety in Residential Care Premises.* The Institute of Building Control 1997.
6. An area in a residential care building with 30 minutets fire protection which helps to reduce travel distances by providing a temporary refuge.
7. *Firecode. HTM 88. Guide to fire precautions in NHS housing in the community for mentally handicapped (or mentally ill) people* (DHSS) HMSO 1986.
8. The Education (Schools and Further and Higher Education) (Amendment) Regulations 2001 (SI 2001/692).
9. BS 5588: Fire precautions in the design, construction and use of buildings: Part 6 : 1991 Code of practice for assembly buildings.

10 BS 5588: Fire precautions in the design, construction and use of buildings: Part 10 : 1991 Code of practice for shopping complexes.
11 BS 5588: Fire precautions in the design, construction and use of buildings: Part 7 : 1997 Code of practice for the incorporation of atria in buildings.
12 BS 5588: Fire precautions in the design, construction and use of buildings: Part 8 : 1999 Code of practice for means of escape for disabled people.
13 Approved Document M, Access and facilities for disabled people. DTLR and the Welsh Office 1999.
14 BS 5810: 1979 Code of practice for access for the disabled to buildings.
15 BS 5655: Lifts and service lifts Part 1 : 1986 Safety rules for the construction and installation of electric lifts and Part 2: 1988 Safety rules for the construction and installation of hydraulic lifts.

Chapter 8

Modification of the Basic Principles of Means of Escape

8.1 Introduction

The previous chapters described the basis on which the means of escape are designed, as laid down in the Building Regulations Approved Document B and various British Standards. There are many reasons why it can be difficult to design fully in accordance with the prescriptive guidance. This chapter describes how the safety objectives can be satisfied by modifying the basic principles, or by taking account of factors not included in the prescriptive guidance.

8.1.1 Guidelines and limits

The most important point to understand is that the various numbers set in guidance documents as limits should not be regarded as pass/fail criteria never to be exceeded. Nor should it always be assumed that staying within the limits is a certain guarantee of adequate safety. There is so much conservatism in the prescribed solutions that it would be unusual for them to lead to an unsafe design, but the possibility should be recognised. More often than not, there are good reasons why small differences will have no significance for the safety of the design.

8.1.2 Means of escape: structural and otherwise

The definition of 'means of escape', quoted in Chapter 1, uses the phrase '*structural means* [my emphasis] whereby a safe route or routes is or are provided...'. This reflects a traditional approach that has been based largely on construction and layout to achieve a degree of protection. Some of the alternative or modified approaches discussed in this chapter employ methods that are based on neither construction nor layout.

8.2 The evacuation process

The prescriptive guidance uses a 'starting gun' analogy for the evacuation process. At a certain point after a fire has started, which is not defined, a warning is given and occupants move more or less directly to exits. The means of escape are designed to cater for the resulting flow of people. Developments in fire safety engineering have led to a different understanding of the process. This is discussed more fully in Chapter 10. For the present, it is enough to say that the occupants' response to signs of a fire is much more complicated than the 'starting gun' concept.

8.2.1 *Effect of behaviour on peak capacity*

One obvious result of the discrepancy between the model used in the standard guidance, and actual behaviour patterns, is that the peak flows in the escape routes are likely to be significantly lower than assumed for sizing those routes. In most situations, the movement of occupants to the exits is spread over time. Contrary to the picture sometimes painted in films or newspapers of a frigh-tened stampede for exits, the difficulty can be getting the occupants to stop what they are doing and leave. Clearly, if no one moves until the last moment crowding is likely, and a number of fatal incidents have occurred in this way. However, unless there is reason to expect occupants to behave as a crowd, the natural range of personal characteristics in the average population leads to a staggered response.

An awareness of this tendency can provide some reassurance if there is a need to depart from normal exit capacity standards.

8.3 Is escape really the right move?

Before considering how the principles of means of escape design can be modified, it is worth asking the question 'is it necessary to have a means of escape?'. For a few specialised applications the best strategy may be to defend in place. This usually only applies in premises where some essential function *has* to be maintained and the consequences of abandoning operations cannot be tolerated. Examples include some process control buildings, and places where the external environment could be hostile. The solution of such special problems is likely to involve fire safety engineering, but the importance of asking this sort of fundamental question still applies in less extreme situations.

Flats are designed on the assumption that it is only the occupants of the flat where a fire has started who will need to evacuate at first, possibly with a few of their immediate neighbours thereafter. On floors above and below the fire, occupants are protected by the construction between each dwelling. In part this strategy developed because little reliance could be placed on general evacuation of this sort of building without outside intervention. But if the principle is good enough for millions of people to live by, why not apply it to other kinds of multi-cellular buildings where people sleep?

8.4 Strategies

Two kinds of modification can be made to improve the means of escape system: they either speed up the evacuation process or slow down the onset of dangerous conditions from the fire. These two simple categories of measures are broken down into more detail in Tables 8.1 and 8.2. The rest of this chapter looks at some of the measures and their implications.

8.5 Basic data on movement in escape routes

The speed of movement varies with the density of people in the escape route. If the density is high, more than four people per square metre, conditions become very uncomfortable and movement is difficult/impossible. On a level surface walking speed can be taken as 1.2 m/s if the density is at least (i.e. no more crowded than) 0.5 m^2/person. Wheelchair users can move at least as quickly on the flat, provided that any doorways are designed in accordance with recommendations in Approved Document M. However, people using walking frames move much more slowly; a design figure of 0.6 m/s is suggested in CIBSE *Guide E*[1].

On stairs the speed of movement is influenced by density of people in the escape route, but the pitch is also a factor. If working near to the maximum pitch acceptable under Approved Document K, the design speed can be taken as 1.1 m/s, giving a vertical velocity component of about 0.75 m/s.

8.6 Balancing exit capacity and travel distance

The early design stages of the means of escape are mainly concerned with establishing the location, number and size of exits and stairways, as described in

Table 8.1 A summary of strategies to reduce the time occupants take to get to safety.

1. Occupants to respond more quickly	Visibility of source of fire	Open plan better than cellular; one possible advantage of atria over separated floors
	Early warning by automatic detection	Prescription often assumes no auto detection
	Voice alarm instead of plain sounder	See section 8.7
	Live alarm messages rather than pre-recorded ones	See section 8.7
	Train occupants or some of occupants (staff)	See section 8.7
2. Occupants to move more quickly	Occupant profile more active than average	E.g. young or sporty population
	Wider circulation routes	Reduces density in route
	Lower occupant density	If density is lower in circulation route
	Arrange assistance	Trained staff
	Avoid impediments	Shallower stairs, avoid uneven surfaces or irregular steps, signal changes in level, lighting
	Provide support	Handrails, barriers
	Mechanical aids to movement	Evacuation lift, moving walkway
3. Improve efficiency of use	Prioritise use so that those in most immediate risk are evacuated first	Phased evacuation, progressive horizontal evacuation
4. Reduce distance to move	Protected routes	Provides higher confidence that part of route will remain usable
	Introduce sub-compartments	Place of relative safety nearby; progressive horizontal evacuation
	Refuges	Delays the need to use stair, if that requires assistance
	Layout changes	
5. Simplify wayfinding	Layout change	Improve orientation cues, reduce number of changes of direction; make escape routes familiar as part of normal circulation
	Interior design	Improve sightlines, increase awareness of surroundings
	Signage	Control competing elements, integrated system of signage

the previous chapters. The prime influences are the number of people and the geometry of the floorplate. The population size determines the capacity and therefore width of the exits, and the geometry determines the number of exits according to the travel distance limits set for the type of occupancy. Underlying this is a simple flow model which is aimed at getting occupants either to safety outside, or to relative safety in a protected space such as a stairway, in a reasonably short time.

Table 8.2 A summary of stratagies to delay or prevent the onset of dangerous conditions.

1. Contain fire/ smoke	Locate fire hazards away from vulnerable areas	
	Enclose hazard at source: box/housing/ room	
	Pressure differentials; keep smoke in or out of certain areas	Needs air path from pressurised space to open air
	Sub-divide floor into sub-compartments	
	Make floors compartment floors	
2. Smoke management	Maintain clear layer: steady state or time limited	Smoke above heads of occupants during escape
	Dilution: for temperature control (avoiding flashover or glass breakage) or visibility	Enable non-fire resisting construction to contain smoke; allow escape through very thin smoke
4. Limited fire growth/size	Control fire load contents and linings	Keep concentrations of combustibles small and separate
	Ventilate	Avoid flashover, reduce heat stress on compartment
	Fixed extinguishing systems	Sprinklers cool combustion products and contain fire, if not extinguish it; other systems act mainly on the combustion reactions

At this simple level it is possible to rebalance the equation between exit capacity and population to allow for longer or shorter travel distances. For example, if the exits are wider than they need to be for the number of people served, then it could be reasonable to extend travel distance. This would be on the basis that the time to clear the space would be the same. The difference between the way this solution works, and the prescriptive model, is that some occupants spend more time travelling to an exit and less queuing to get through it. Conversely, if the exits cannot be as wide as prescribed for the design population numbers, but the travel distances are shorter than prescribed, then it might still be possible to show that the overall time to clear was much the same.

8.6.1 Density of occupation

The number of occupants of an area can be determined from the design requirement. The authorities are likely to compare these figures with those given as guides in the codes and Approved Document B, but as explained at the start of this chapter and in the AD B, it is acceptable to use other figures if they can be justified.

In Chapter 9 the density of population is introduced into the travel distance/exit capacity equation. With a low density, travel distance is likely to be the factor that determines the time to clear a space. When the density is high, the key factor is exit capacity, to deal with queuing at the exits (see Fig. 8.1). The change from one regime to the other happens at a density of around $4\,m^2$/person.

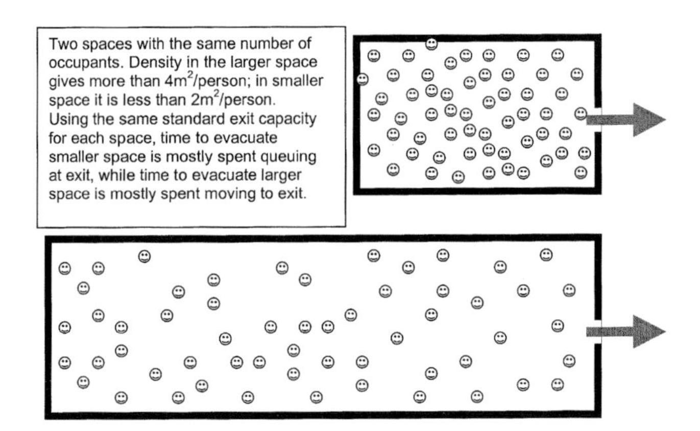

Fig. 8.1 Density of occupation determines whether time taken travelling to an exit, or time spent queuing at an exit, governs overall time to escape.

There are limits to the extent of such trading off between exit width and travel distance. There has to be confidence that, with an extended travel distance, the risk of the route to the exits being blocked by the fire or its products of combustion has not been overlooked.

8.6.2 The time factor in prescriptive guidance on escape

When the flow models were developed, in the first part of the twentieth century, there was less understanding of how long it might take for dangerous conditions to develop. It can be inferred from the flow rates used that the expectation was to clear a space in $2\frac{1}{2}$–3 minutes. But there is little or no connection between this period and the time periods associated with concepts like ASET (available safe escape time) and premovement time, used in fire safety engineering analysis. The fire safety engineering approach, which aims to take account of human behaviour in a more realistic way, is discussed in Chapter 10.

8.7 Improving the occupants' response to fire warning

Fatal fires at department stores in Manchester and in Chesterfield showed how reluctant people can be to interpret and react to signs of danger, especially if engrossed in some activity. In those fires some people eating in the store café took too long to decide to abandon their meal, even though the fire was relatively close to them. Other examples of delayed response have been blamed for many multiple-fatality incidents.

Prescriptive guidance has been more concerned with providing adequate exit capacity from rooms and storeys, than with arrangements to initiate evacuation. Early public health fire safety requirements were aimed at places of assembly, which had a history of multiple fatality fires, and concern that overcrowding in evacuation might block exits.

8.7.1 *Effect of different types of warning*

In the mid 1980s the Fire Research Station of the Building Research Establishment commissioned research[2] into the effect of different kinds of fire warning system on the response of building occupants. This work showed that there could be a very large difference between the response evoked by traditional alarm bells and a spoken warning message through a public address system. In the worst case the traditional sounder produced no useful response at all in some of the experimental subjects. The experimental set-up involved small and isolated groups of people who had no particular affiliation and no prior knowledge either of the building or of the impending fire warning signal. It therefore represented a relatively difficult population to motivate.

8.7.2 *Other psychological effects*

Other psychological studies have shown that people are predisposed not to react to an warning signal, especially if there is some ambiguity in the warning. There are differences between individuals in this respect, which depend partly on the role they are playing at the time. If they have a responsibility for others in the building, whether as a parent with children or as a manager in an organisation's hierarchy, they are likely to respond more quickly.

The spoken message is less ambiguous than a bell or sounder, which the observer can easily decide is not a fire warning, or not for their part of the building, or is a mistake. However, if a spoken message is pre-recorded it can have less impact than a live announcement. This is not normally considered to

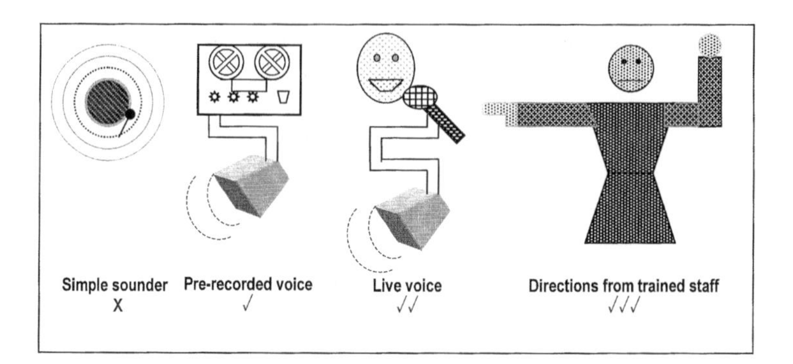

Fig. 8.2 Hierarchy of the effectiveness of fire warning systems, in terms of their ability to motivate a prompt response among the target population.

be vital. Where the best possible response is necessary, the use of live announcements is preferred. The running cost of this arrangement must include trained personnel to staff a control room whenever the building is occupied, and it is therefore only likely to be appropriate in premises with a full-time management team.

It therefore follows that the provision of a higher standard of fire warning system than is called for by the prescriptive guidance, may justify a relaxation of the normal standard for features such as travel distance or exit capacity.

It is not usually possible for the designer to do anything to affect the characteristics of the occupants of a building. But if the designer can establish that the occupants will have some attributes that tend to improve their response in an emergency, it can be useful. For example, training can have significant benefits, and if the building is to be occupied by a company with an established record in this respect, it may be possible to argue that the response to a warning will be quicker.

8.7.3 Role of trained staff in evacuation

The University of Ulster have carried out monitored evacuation tests for the Fire Research Station, from some large shops. These showed that trained staff could largely overcome the natural inertia of customers, with the result that there was little delay in the initial response. In one of these experiments a very large single storey supermarket was emptied of customers in about $1\frac{1}{4}$ minutes. However, it can be difficult to demonstrate to the regulatory authorities that a new building will be managed, and have suitably trained staff, on a long-term basis and with a reasonable degree of certainty. It may be easier to do this with existing buildings if the occupants have already established a track record.

Similarly if a new building is to be occupied by a concern whose existing premises can show good safety management, some account may be taken of this factor in a risk assessment.

8.8 Providing an earlier warning of fire

There are relatively few situations where the guidance in Approved Document B or the BS 5588 series recommends the installation of an automatic fire detection system. Normally it is considered sufficient to rely on the people in the building to detect and give warning of a fire.

However, although our senses are very good at detecting signs of fire, people are not necessarily good at interpreting those signs correctly and taking the most appropriate action. Not even when flames are visible is there a guarantee that the observer will do anything about it. This was graphically demonstrated by a security surveillance video of a shop where people continued to go about their business while a substantial fire developed beside them. When the signals are more ambiguous, as they might be with smells or sounds coming from a fire that is out of sight, the tendency to ignore them is greater. Once people have begun to suspect that something is wrong, they are more likely to go and investigate before taking the 'radical' step of initiating a fire alarm, which slows the start of evacuation.

8.8.1 Automatic detection

An automatic detection system connected to the fire warning system can avoid or at least reduce these delays, although occupants may still seek confirmation for themselves. It ought therefore to be assumed that more time is available to evacuate the building. This extra time can logically be translated into longer travel distance, or can be used to allow for some other departure from normal practice.

The benefit should be greatest in buildings where the reliability of the occupants is lowest. Examples include places where:

- people sleep;
- a significant proportion of the occupants are highly dependent on others, or have limited mobility;
- occupants are unfamiliar with the building;
- occupants are not part of a managed, organised or disciplined group;
- the everyday environment (e.g. noise or smell) might disguise early signs of a fire.

If the occupants may be asleep or incapacitated in some other way, as in a hotel or hospital, standard guidance includes a recommendation to install automatic detection, at least covering the main escape circulation routes. A simple strategy is to specify a higher level of automatic detection, by installing sensors more comprehensively so that all rooms and all unoccupied spaces, such as storage and plant rooms, are covered.

The problem of unwanted alarms can undermine the value of automatic fire detection, as an asset to trade off against other aspects of the design. More than a very few unnecessary evacuations will tend to make the occupants take any fire warning less seriously. However, detector system technology is becoming steadily more reliable. The analogue addressable systems now in quite common use have software that monitors the condition of each sensor, and should eliminate certain types of spurious alarm. Some manufacturers have recently started to offer guarantees against unwanted alarms, although somewhat circumscribed ones. While the actual value of these undertakings may be limited at present, it is a hopeful sign that reliability will go on improving. This should reduce the concern that occupants will assume that an alarm is false and may not react to it.

8.8.2 *Detection to initiate other actions*

Besides giving a warning, automatic detection opens up a range of other possible actions to improve the level of protection (Fig. 8.3):

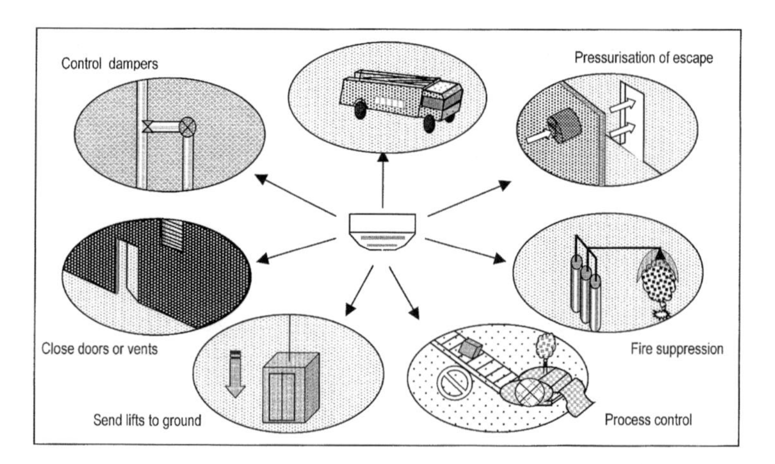

Fig. 8.3 Initiation of fire protection measures on signal from automatic detection.

- sending a warning off-site to summon assistance:
  - besides calling the fire service, interfaces with radio paging devices can call key personnel, security patrols, etc.;
- closing down recirculatory ventilation to avoid smoke spread:
  - generally called for under standard guidance in any case;
- causing ventilation equipment to create pressure differentials to affect smoke movement:
  - positive or negative pressure to keep smoke out of certain areas, within a floor or from one floor to another;
- operating fire suppression systems, opening smoke vents:
  - beware of undesirable interactions between suppression and heat/smoke ventilation; the operation of one system may be delayed by the working of the other;
- closing fire doors, dampers and shutters:
  - allowing relatively open spaces, in normal circumstances, to be subdivided against smoke or fire spread;
- stopping process equipment:
  - to remove dangers associated with unsupervised operation, without delaying people who would otherwise have to stop equipment, and to encourage occupants to leave by withdrawing services or activities they were engaged with;
- instructing lift control equipment to initiate homing routines.

8.9 Mechanical aids to movement

The use of escalators on escape routes has been ruled out in most building codes, despite the fact that most of the world's urban mass transit systems depend heavily on them (and their associated codes presume that they will be used in emergencies). There are two reasons why they are ruled out. The first perceived problem is associated with concern about falls, and the pile-up that might result:

- the ratio of rise and going is not ideal for rapid movement if the escalator is stopped (and it varies at top and bottom as the steps flatten out);
- if the escalator stops abruptly when people are on it, they may fall;
- there may be many more risers in an uninterrupted flight than allowed for in a normal stair.

These problems could arise in everyday use and can be mitigated by a secure power supply. Soft stop machines would reduce the danger of overbalancing, but they increase the danger should a person be caught in the mechanism.

The second perceived risk is that the equipment will not be available when needed. This may be because it is undergoing maintenance; or because of a power failure, either in the public supply or as a result of the fire. Fire in the machinery is unlikely to cause the immediate failure of the escalator if the motors are at the bottom. The question of maintenance is probably the most intractable of the failure modes since, over the life of the equipment, it accounts for much more downtime than the other faults. It may not matter where there are three escalators to cater for tidal flows, or if there are two and there is scope for reversing the available unit in an emergency. The fallback position is to assume the use of the escalator as a static stairway, although the objection to the variation in the rise:going ratio remains.

It is also necessary to think about the protection of the space around an escalator. To be on an equal footing with a protected stairway, the escalators would be expected to be in a protected enclosure. The experience of the Kings Cross tube station disaster in London, where the fire is thought to have begun in, and spread via, the materials of the escalator, should not rule out the use of escalators for escape. Modern machines can be constructed mainly of non-combustible material, and operators are aware of the need to clear combustible debris from the machine spaces.

If these precautions are taken it may be possible to include the escalator as part of the means of escape. In estimating the flow rate of people on them it is conservative to assume that they are stopped and to treat them as an ordinary stair.

8.9.1 *Lifts*

New buildings in the UK over 18 m in height will have fire-fighting lifts, which are designed to be suitable for evacuation of disabled people, and for use by fire-fighters during a fire. Such lifts are not common in other countries although interest is increasing, and research is going on in a few other places. Although these lifts may be relatively fast, because they should reach the top storey in about 60 seconds, they are generally small (eight person) passenger lifts, and therefore do not have significant capacity for the general evacuation of the building.

The next step will be to capitalise on the much greater capacity of the ordinary passenger lifts in tall buildings, and find ways of using them to evacuate the main mass of occupants. To deal with the normal morning rush-hour peak, lifts in high-rise offices can move 15% or more of the building's population in five minutes.

8.10 The benefits of fire suppression systems

Sprinklers are the most common automatic suppression systems, and this section will concentrate on them. They have a very good record for reliability and effectiveness. It is very unusual for a fire to cause fatal injuries in a sprinklered building.

8.10.1 Sprinklers

A sprinkler head takes longer to respond to the effects of a fire than a smoke detector, but quick response sprinkler heads are similar to heat detectors. They react when the heat release rate from a fast growing fire is about 1 MW (with some variation according to the height of the room). A slower fire growth rate will lead to activation at a lower heat release rate.

Sprinklers' effect on safety in room of origin and elsewhere

Despite this improved technology, by the time the sprinkler head operates, conditions in a room where the fire has started could be dangerous, unless the volume of the room is particularly large. Therefore a sprinkler system will not necessarily be of much benefit to the occupants of the room of fire origin in the case of a hotel bedroom or equivalent. They will probably have to make their escape before the sprinkler has operated, or risk being overcome by the toxic products of combustion. However in larger spaces, and in rooms other than the room of origin, sprinklers can have a very significant effect on the conditions the occupants face.

Sprinkler action

The action of the sprinkler system is to slow down the spread of fire by wetting the surrounding area, as well as to cool the seat of the fire and the smoke and gas rising from it. Water also has an inhibiting effect on the combustion process, in the gas phase. (See Fig. 8.4) Sometimes obstructions, such as shelves or other furniture, shield the seat of the fire from the water. If it does not extinguish the fire completely, the sprinkler prevents it from getting larger, and makes the job of the fire service easier.

An unsprinklered fire causes a layer of very hot gas to spread under the ceiling. Downward radiation from this layer preheats surfaces in the room, causing ever more rapid growth of the burning area. This 'chain reaction' can lead to a point where all the surfaces in the room are ignited. Temperature rises rapidly with the risk of break-out into other rooms or corridors. With rising

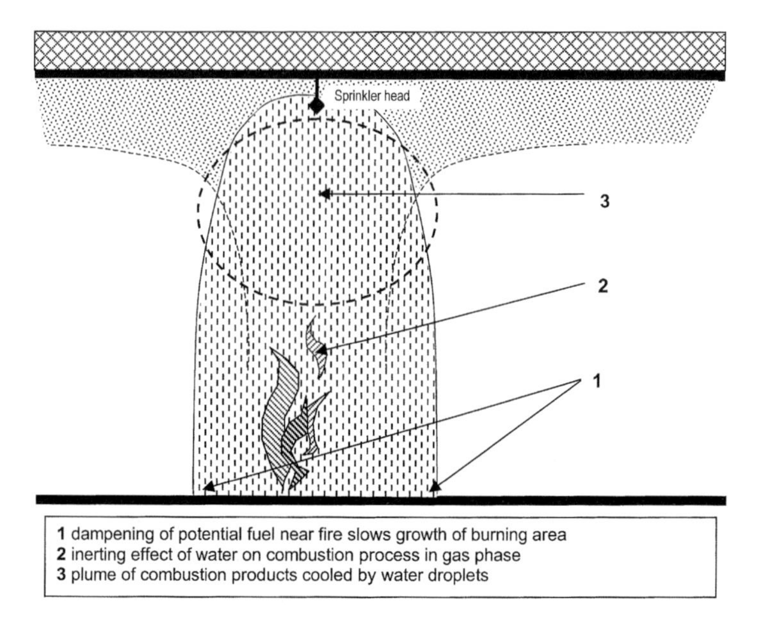

Fig. 8.4 Sprinkler action.

temperature come significant pressure differences between the room and the surroundings which can drive spread of very large quantities of dangerous smoke and gas. A very high proportion of sprinkler system fires lead to no more than four heads operating. This indicates that sprinklers usually stop the very hot ceiling layer from spreading more than a few metres in any direction, and prevent the 'chain reaction' from getting under way.

Sprinklers and glazed partitions

The ability of sprinklers to keep the average smoke layer temperature down to around 100°C can be exploited. For example, ordinary toughened or laminated glass can be used in construction designed to separate occupied spaces from the effects of fire, instead of more expensive fire rated glass.

Trade-off between sprinklers and other escape route design features

It is generally agreed that if the sprinklers perform, the fire is of little danger outside the room of origin. Logically it should therefore be possible to dispense with most other fire protection measures when a properly designed and installed sprinkler system is present. In practice the concessions are not so dramatic. The reason usually given is that sprinklers are not completely reliable, and other lines of defence must therefore still be maintained. The often-heard comment when discussing sprinkler-based relaxation of some prescribed solution is 'what happens if the sprinklers don't work?'. This attitude is not consistent with the less critical acceptance of passive fire protection measures, notably fire doors, on which the safety of occupants also depends. One might just as well consider 'what happens if the fire doors don't work?' but it is not a particularly useful negotiating point.

Sprinkler reliability

Estimates for the 'reliability' of sprinkler systems vary between over 60% and over 90%, the variation being partly due to different definitions of success or failure. Fire door surveys tend to show that fewer than 50% of the doors would function as intended, mainly because of wear and tear. The use of properly designed hold-open devices can improve door reliability, by removing most of this wear. This is provided that the twisting forces set up between the door closer and the holder do not permanently distort the door and prevent it making a good seal in the opening.

It is easier to think of a single point failure that disables a sprinkler system, than it is to accept that a particular fire door in a critical location will be ineffective. But the probability of the failure event is much smaller for the sprinkler system than for a fire door. In a life safety sprinkler system the potential for single point failure, such as water supply being turned off, has been addressed, for example by having monitored valves normally locked open.

Summary of sprinkler effects

When considering the benefits of sprinklers to modify the principles of a conventional escape strategy:

- conditions in the room of origin will be largely unaffected by sprinklers while the room is being evacuated, unless the room is large, say over 100 m^2 (a very approximate guide) with at least a 3 m ceiling height;
- in large rooms sprinklers can delay the onset of dangerous conditions;
- conditions outside the room of origin will be improved by sprinklers in the room of origin, and in nearly every case will present little risk to occupants;
- significant relaxations of travel distance or other code recommendations for escape route design are unlikely to be approved as a trade-off against sprinklers without carrying out a fire engineering analysis.

8.11 Dealing with smoke

There are various ways of preventing smoke from inhibiting escape:

- containment by solid barriers, or by using pressure differentials and the air flows these set up, to keep the smoke out of prime escape routes such as stairways;
- controlling and directing smoke flow, using its buoyancy, channelling screens and ducts to take it to a place (usually the open air) where it is relatively harmless;
- dilution for temperature control or tenability;
- using fire suppression to control the production rate.

There is also the issue of smoke clearance which, though strictly nothing to do with escape, is discussed at the end of this section because of confusion that sometimes arises over its role.

Implementation of these principles is likely to involve fire safety engineering, and more will be said about them in Chapter 10. The following points are intended to help the reader decide whether one of these methods might be useful in a particular situation.

8.11.1 Smoke containment

The difficulty with containment as an approach is that the architectural and functional requirement is often for openness rather than enclosure (Fig. 8.5). Pressurisation is not necessarily any advantage in this respect because it also depends on physical barriers between the higher and lower pressure areas (Fig. 8.6). Recent code changes have also made it considerably more complicated to construct and commission pressurisation systems without interfering with façade design and taking up floor area for duct space.

8.11.2 Existing buildings: considerations with pressurisation

In existing buildings, particularly though not exclusively pre-twentieth century ones, the reliability of the main elements of construction as barriers to smoke or fire is an issue. If an adequate estimate of leakage areas can be made, pressurisation, say of a stairway, can be a less damaging way of making it a 'place of relative safety' than upgrading the construction to a fire-resisting standard. This approach still presents challenges, such as finding suitable air inlet locations, minimising the impact of ducts, and securing a power supply. The

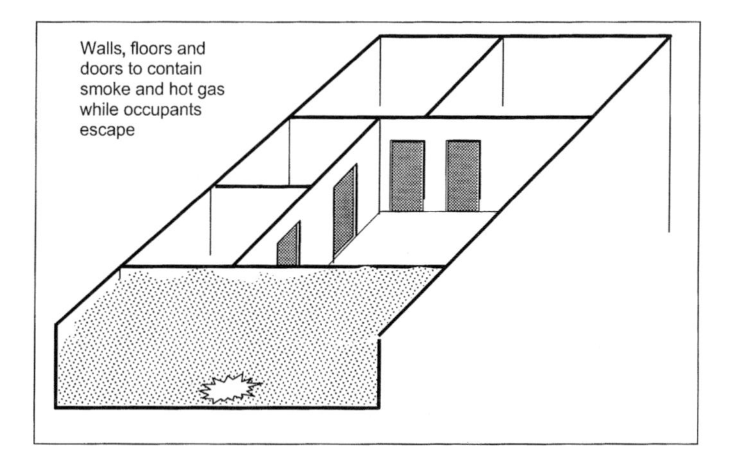

Fig. 8.5 Smoke movement limited by sub-dividing construction.

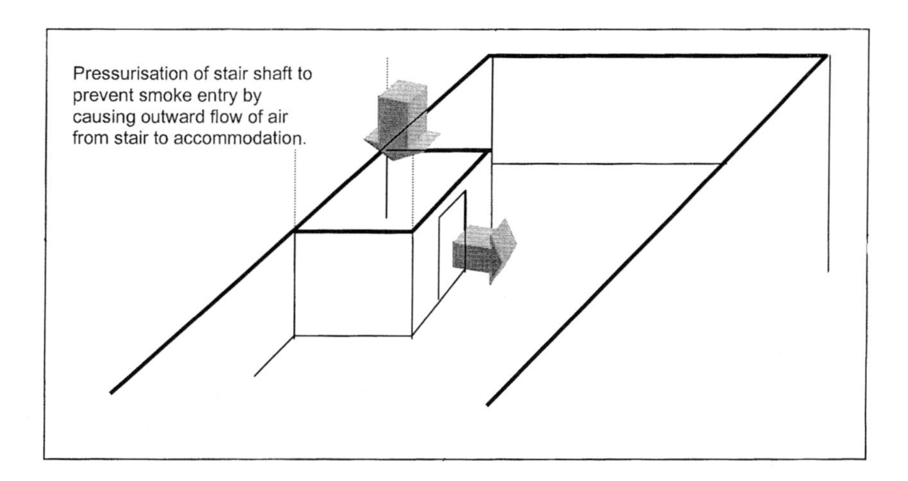

Fig. 8.6 Smoke control by pressure differences between protected space and accommodation.

commissioning process can be lengthy because of the uncertainty about leakage paths.

8.11.3 *The cabin concept*

In large spaces where the fire loads are localised and fixed, the 'cabin' concept can be used to contain the products of combustion without breaking up the space with many walls. The principle is similar to that of a cooker hood. A reservoir is formed above the area where the fire load is concentrated, and some arrangement is made for removing smoke that collects there so that it does not spread any further (Fig. 8.7). People in the area around the reservoir are then in no immediate danger. Sprinklers are normally installed in the 'cabin' so that the fire size can be limited. This makes it easier to predict the rate at which smoke will have to be extracted from the reservoir to prevent it overfilling.

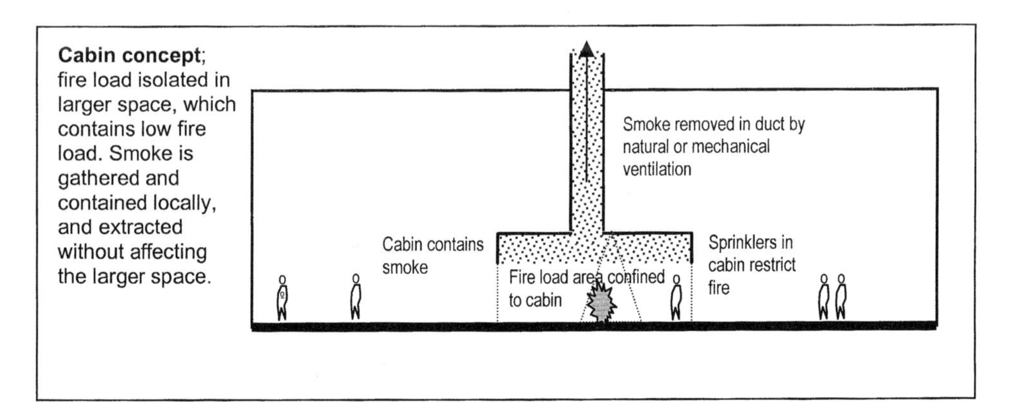

Fig. 8.7 The cabin concept.

The cabin concept is not usually applicable to older buildings. The idea was first used in airport concourses, but it has also been used in the modernisation of facilities in some historic railway stations.

8.11.4 Shopping mall smoke control methods

The cabin concept was an extension of the smoke control principles developed for covered shopping centres. There, instead of extracting smoke at source, a common route is provided that leads smoke from any shop to vents or mechanical extract points. The route is usually the same circulation corridor or mall that occupants use for escape. The buoyancy of the hot smoke, and the design of channelling screens and downstands used to confine it, keeps it safely above people's heads, while it is ventilated to the atmosphere at roof level (Fig. 8.8).

There is an alternative approach which keeps smoke out of the main mall by extracting directly into ducts, from the shop unit where the fire occurs. This tends to require more complicated control systems, with motorised dampers to isolate all but the unit on fire from the extract ductwork. But there are situations, such as a mall with a low ceiling, where this is the better approach.

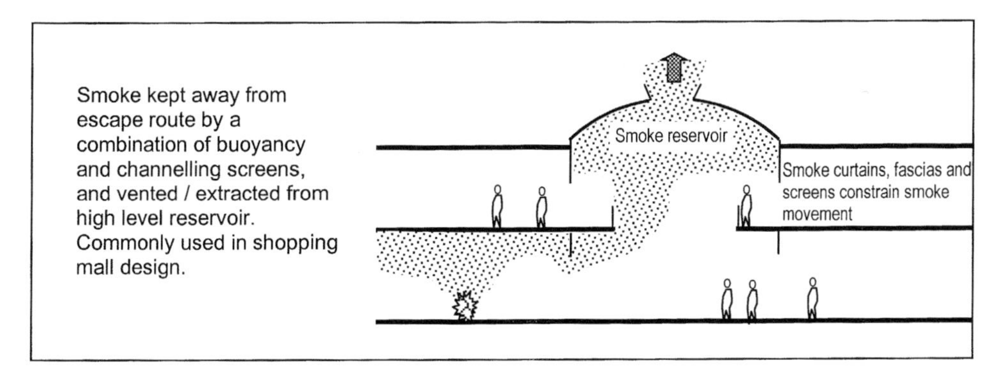

Fig. 8.8 Typical smoke control of a shopping mall using natural ventilation.

8.11.5 Design fires

In any of these schemes the temperature and volume of the smoke are important factors. It is possible to estimate these factors if enough is known about the fire load. But the fire load is often either only roughly known or subject to change in the future. In this case sprinklers can be helpful. The presence of sprinklers usually allows a maximum fire size to be determined and used in the smoke control system design process.

8.11.6 Smoke dilution

Mixing smoke with cool fresh air reduces the average temperature of the smoke layer and the concentration of the harmful or irritating components. It is usually impractical to supply enough fresh air to render a smoky atmosphere breathable. However, smoke rising several storeys in a tall atrium will entrain air, and this dilution will cool it very significantly.

Dilution can be used to keep smoke in a space below some critical temperature. This might be to allow people to escape under the smoke layer without being exposed to excessive radiation. It could be to allow the use of ordinary toughened glass in partitions enclosing the space, rather than fire-resisting glazing. The common example is an atrium which is enclosed for acoustic or other environmental reasons. By installing temperature control ventilation a fire-resisting enclosure can be dispensed with, while still allowing a phased evacuation to be used in the accommodation adjoining the atrium.

The technique is normally to ensure that there is sufficient outlet ventilation (with matching low level inlets) to keep a smoke layer base at a certain height above the highest point that smoke could enter the space. This ensures that there will be enough air entrainment to cool the smoke plume below the critical temperature (between 200°C and 400°C in the case of glass breakage). The system can use fans, or be based on natural buoyancy.

More is said about design temperature criteria of smoke control systems in Chapter 10.

8.11.7 Smoke clearance ventilation

Smoke clearance ventilation is asked for in some codes, e.g. BS 5588 : Part 7, and the guidance on the application of section 20 of the London Building Acts in Inner London (LDSA Guide 1[3]). In Part 7 it is reasonably clear that the purpose is to get rid of relatively cool smoke after a fire. It is known that the BSI committee was aware of the problem (exemplified by a fire at the World Bank in Washington DC) of clearing smoke from a fully enclosed atrium.

However, as the fire service has developed a better understanding of the role of ventilation in combustion processes, they have become more interested in having some control over it during fire-fighting operations. This is a safety issue for the fire-fighter and therefore a matter of Health and Safety at Work legislation. At present the Approved Document B only recommends heat and smoke ventilation as part of the access and facilities for fire-fighters in basements. It does not require smoke clearance ventilation at all, because post-fire recovery is strictly a property or business protection measure, to reduce smoke damage and speed up the rehabilitation of the building after a fire.

The recent clarifications in the 2000 edition of the Approved Document B, by which many atrium buildings of up to 30 m height will not require compartmentation, and therefore will not, in the terms of the Approved Document, need to follow guidance in BS 5588 : Part 7[4], are likely to highlight the ventilation issue. In Inner London the trend seems to be toward fire ventilation, with systems capable of removing smoke at high temperature.

There is also a technical issue. The rules of thumb for smoke clearance were based on relating vent area to floor area. This may not be appropriate if the aim is to improve the safety of fire-fighters from unburnt gases inside a compartment, rather than flushing cold smoke out of a whole storey in a reasonably short time to permit reoccupation of the space. It is the latter function that was envisaged when code-writers set current rules of thumb.

8.12 References

1 The Chartered Institution of Building Services Engineers. *Guide E: Fire Engineering*. CIBSE, London, 1997.

2 Building Research Establishment Report BR 172: Experimental programme to investigate informative fire warning characteristics for motivating fast evacuation.
3 London District Surveyors' Association: *Fire Safety Guide No. 1, Fire Safety in Section 20 Buildings*. LDSA Publications, Beckenham, Kent, 1997.
4 BS 5588: Fire precautions in the design, construction and use of buildings: Part 7: 1995; Code of practice for the incorporation of atria in buildings.

Chapter 9

New Approaches 1: BS 9999 : Part 1 Means of Escape

9.1 Introduction

In 1998, the BSI commissioned consultants to write the draft of a new British Standard on precautions against fire in buildings. The intention is that the new Standard will not only replace the BS 5588 series, but could fulfil the function of many other fire safety guidance documents that have traditionally been produced by Government departments and other bodies. There are roughly 100 documents of this kind. Some are highly specialised, but there is a good deal of overlap between some of the others, and the recommendations are not always consistent with one another. The sheer volume of material is a problem for practitioners too.

Earlier in the 1990s, there was a review of fire legislation and its enforcement by the Government. This identified the large amount of guidance material as a source of widespread complaint from all sectors of the construction industry, fire protection industry and fire service.

The idea for BS 9999 – to consolidate all the existing guidance into one standard – emerged as one result of that review. It would still be basically a prescriptive document, but would set out principles, based as much as possible on fire engineering analysis, so that it could be applied to a very wide range of building conditions. This original intention has been diluted by the decision not to include hospitals or transport termini in the range of building types covered by the draft BS 9999.

It was decided to divide the document into four parts:

- Part 1 Means of escape;
- Part 2 Construction;
- Part 3 Facilities for the fire service;
- Part 4 Management of fire safety.

The material in this chapter is based on the draft of Part 1: Means of Escape as it went to public consultation in 2001[1]. There may be significant changes before the Standard is published. At the time of writing, a timetable for publication of BS 9999 was not known. From experience of the preparation of other major fire standards, it is quite possible that committee discussion will take several years following the public consultation.

However, there are some interesting proposals in the draft of Part 1 that are worth describing now. There is no reason why some of the ideas should not be used in practice before the standard is published, although it will be much easier to adopt them once they are embodied in a code.

9.2 A time and risk-based approach

The fundamental principle underlying the recommendations of the draft of Part 1 of BS 9999 on means of escape, is a simplified time-line analysis – that is, the process of determining whether occupants can get out of danger before conditions become untenable. The allowable travel time must be equal to or greater than the calculated travel time. If the allowable travel time is exceeded, conditions may become untenable before occupants reach a place of relative safety (see Fig. 9.1).

According to the draft Standard, each building purpose group is assigned to a risk category. The purpose groups are substantially the same as those described in the Building Regulations Approved Document B. A minimum set of fire safety measures is prescribed for each risk category, intended to ensure that the

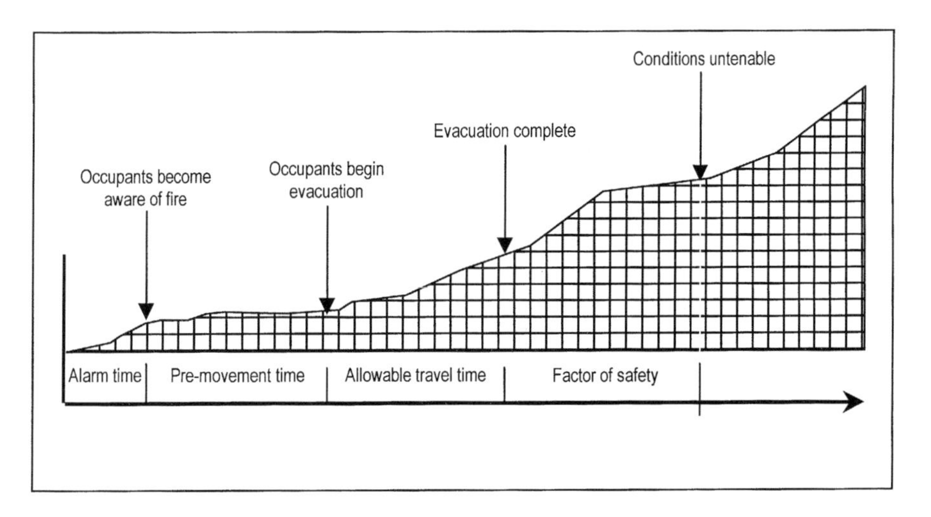

Fig. 9.1 Allowable travel time At_{trav} in relation to stages in the alarm and evacuation process.

allowable travel time will not be exceeded. This minimum set of measures represents a baseline. If extra fire protection measures are taken, the allowable travel time can be increased, up to a prescribed limit, with the result that important factors such as travel distance can be extended.

9.2.1 *Risk categories*

The classification of risk is based on occupancy profile and fire potential.

The occupancy profile is a broad categorisation of the sort of people who can be expected to make up most of the occupants (Table 9.1). The categories used are very much the same as in BS 5588 : Part 7, the atrium code[2].

Table 9.1 Occupancy profiles.

Category	Profile	Examples
A	Awake and familiar with building	Office, factory, teaching area of school
B	Awake but unfamiliar with building	Shop, museum, assembly building
C	Asleep:	
	ci- Long term individual occupancy	Flats without 24-hour management control of maintenance of the premises
	cii- Long term managed occupancy	Serviced flats, hall of residence, sleeping area in boarding school
	ciii- Short term occupancy	Hotel
D	Medical care	Hospital
E	Transport terminal	Railway station, airport passenger terminal

Note: Group D and E buildings are not within the scope of the draft BS 9999 Standards as currently proposed.

Fire potential represents both the fire load, a measure of the amount of combustible material in the building, and factors such as its disposition that affect the speed with which a fire will grow (Table 9.2).

Table 9.2 Fire potential.

Class	Fire growth rate	Examples
1	Slow	Picture gallery
2	Medium	Dwelling, office, hospital, hotel reception or hotel bedroom
3	Fast	Shop
4	Ultra fast	Industrial storage or plant

9.3 Minimum fire protection measures

These measures represent the baseline. The draft standard expects them to be provided in every case.

For all occupancies

- a fire alarm system;
- automatic detection where phased, staged or zoned evacuation is involved;
- escape lighting;
- fire safety management (covered in Part 4 of BS 9999);
- exit route signage.

There is also guidance on:

- doors and final exits:
  - where density is high (less than 4 m^2/person) the aggregate exit width needed from a space is calculated from: number of people + (allowable travel time)$^{0.9}$;
  - escape routes should be at least 1200 mm wide, unless the area served is not accessible to people in wheelchairs and has a population of fewer than 60 people;
  - doorways on escape routes can be 150 mm narrower than the route;
- the enclosure of higher fire risk areas;
- the use of protected power circuits;
- lifts, including evacuation lifts;
- mechanical ventilation systems;
- refuse handling.

For category A (see Table 9.1) occupancies, in addition to the minimum provision

- fire alarm system;
- escape lighting.

For category B occupancies, in addition to the minimum provision

- fire alarm system;
- automatic fire detection for large or complex premises (population of over 300 persons), or for areas only visited occasionally;
- escape lighting.

For category C occupancies, in addition to the minimum provision

- automatic fire detection and alarm system;
- escape lighting.

For category D occupancies, in addition to the minimum provision

- (refer to *Firecode* – the series of guides produced by NHS Estates on aspects of fire safety in health care premises[3]).

9.4 Travel distance and time

In spaces where the density of occupants is low (where there are more than 4 m^2/person) the time taken to leave the space depends on the travel distance. In more crowded spaces the determinant is the queuing time at the exits (Fig. 9.2).

The allowable travel time includes both walking and queuing time involved in getting to a place of relative safety. Conservative estimates have been made for the pre-movement time, detection/alarm time, when the allowable travel

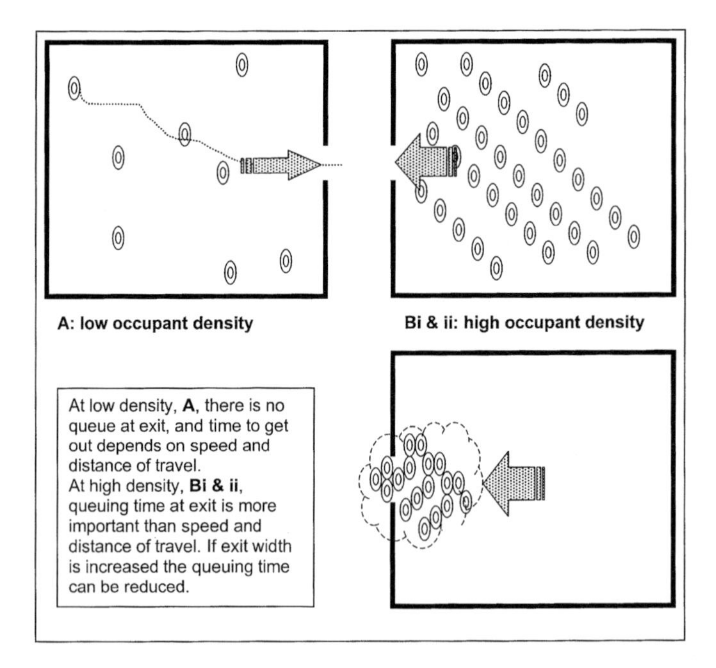

Fig. 9.2 Relationship between occupant density, travel distance and exit capacity.

Table 9.3 Maximum allowable travel time, in minutes, (MATT) by risk category.

Risk category	MATT with minimum set of protection measures	MATT with extra fire protection measures	MATT for dead-end in a room
A1	3.8	5.7	1.1
A2	3.0	4.5	0.7
A3	2.2	3.3	0.7
A4	1.3	1.9	0.4
B1	3.7	5.5	1.1
B2	2.7	4.0	0.8
B3	1.7	2.5	0.5
B4	Not allowed	Not allowed	Not allowed
C1	3.7	5.5	1.1
C2	2.7	4.0	0.8
C3	1.7	2.5	0.5
C4	Not allowed	Not allowed	Not allowed
D E	Not within the scope of the guidance in BS 9999		

Note: 'Not allowed' indicates that extra protection measures are needed which will bring the fire potential of this occupancy down. An automatic sprinkler installation would do this.

time limits were being determined. The draft Standard gives a range of allowable travel time for all the combinations of occupant risk and fire potential (see Table 9.3).

9.4.1 *Variations to ATT*

The basic travel times can be increased, up to specified limit values, if extra protection measures are provided. The basic travel times may also have to be reduced in some cases; for example, where alcohol is consumed on the premises, it is recommended that the ATT is reduced by 25%.

Sprinklers

The draft proposes that the incorporation of a sprinkler system could allow the risk category to be reduced by one grade; e.g. A4 would become A3, or B4 would become B3, instead of being 'not allowed'.

High ceilings

Allowance can be made for high ceiling rooms, in premises of over 200 m^2, in Occupant Risk categories A, B and E. This is because the smoke will take longer to build down to head height, under a high ceiling.

9.4.2 *Travel speed and distance*

Low occupant density

In occupant risk category A, a travel speed of 0.4 m/s has been proposed. For the other categories it is 0.3 m/s. These are very low speeds, to take account of the whole range of complicating factors that can make the evacuation process much less efficient in practice than simple observation might suggest. Despite this conservatism, when these speeds are applied to the allowable travel time the maximum travel distances can be considerably longer than current guidance recommends (Table 9.4).

Table 9.4 Examples of travel distances for low density of occupation.

Risk category	Max travel (m): minimum set of precautions	Max travel (m): top limit not to be exceeded when extra measures are added	Dead-end travel distance (m)
A1	91	136	27
A4	31	46	9
B1	66	99	20
B3	30	45	9
C1	66	99	20
C3	30	45	9

High density occupation

The draft Standard proposes that the same limits should be applied in high density occupancies. Note that the current 5 mm of exit width/occupant type of rule is replaced in the draft by the relationship:

$$\text{Aggregate exit width} = \text{number of people} + (\text{allowable travel time})^{0.9}$$

It is proposed that the usual rule, that the largest exit is discounted, should still be used.

9.4.3 *Cellular accommodation*

The draft Standard proposes a number of rules for designing the width of corridors and storey exits from cellular accommodation. In many instances the 1200 mm minimum width suggested for all routes will not need to be exceeded, but the point is also made that partitions should be carried to the structural slab, or to a ceiling which is of smoke retarding construction. In buildings where the partitions normally stop at a suspended ceiling, below a heavily serviced ceiling void, this could be a significant additional complication.

9.4.4 *Alternative routes*

The consultants who drafted Part 1 have suggested a simple rule to ensure that two exits are sufficiently far apart to provide a real alternative (Fig. 9.3):

$$\text{Maximum room dimension} \div 2 \leq \text{distance between exits}$$

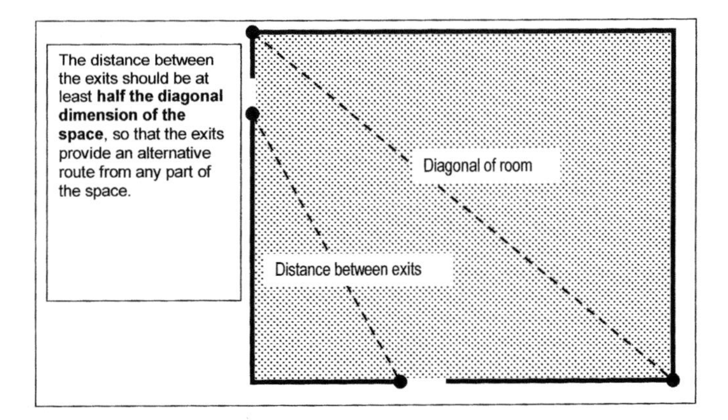

Fig. 9.3 Rule for the separation of exits from open-plan spaces.

9.5 Stairway sizing

The draft carries the idea of allowable travel time into the sizing of protected stairways. It contains graphs of stair width to serve the total simultaneous evacuation of a specified number of occupants, given the number of storeys and the minimum available travel time for the occupants entering the stairway. There is also a formula to enable the width to be calculated for a particular arrangement. A modified version of the formula enables escalators to be sized for escape purposes. This application is limited to a few situations where only a single escalator is used (i.e. not a sequence serving a series of storeys). The basis of these particular proposals is not given, although the work of Fruin and Pauls is mentioned.

The width of a stair in a phased evacuation building can be derived from the same graphs or equation, but using the number of storeys and people in one phase, rather than the total number of storeys and building population.

9.6 Other topics

The draft covers the other aspects of means of escape design in much the same way as current parts of the BS 5588 series. The second half of the BS 9999 : Part 1 draft document is a useful summary of guidance presented on a building-type by building-type basis.

In the section on shopping malls there are new proposals for assessing the occupant capacity of the development. These are based on a relationship between gross lettable area and density, in which the density of occupation in larger developments is lower than in small ones. The information is based on North American data, and will no doubt be scrutinised closely during the consultation process.

In some areas the draft has fallen back on referring to other sources of guidance. This applies particularly to houses in multiple occupation, residential health care and sheltered housing. No doubt these problems will be addressed before the final version is published. As already mentioned, transport termini are currently not within the scope of the draft Standard.

9.7 Conclusion

The draft of BS 9999 : Part 1 on means of escape offers some significant advances over current guidance. The technical issues are well presented and should enable much more flexible or economical designs to be made. Travel distances will be increased and the number of stairways will be reduced in some commercial building types, where such gains will be of great value to developers.

9.8 References

1 British Standards Institution: 2001: draft BS 9999-2; Code of practice for fire safety in the design, construction and use of buildings; Section 6: Means of escape.
2 British Standards Institution; BS 5588: Fire precautions in the design, construction and use of buildings: Part 7: 1995; Code of practice for the incorporation of atria in buildings; BSI London.
3 NHS Estates; 1996 *Firecode*. Health Technical Memorandum 81. Fire precautions in new hospitals. HMSO, London.

Chapter 10
New Approaches 2: Fire Safety Engineering

10.1 Introduction

All the preceding parts of this book have been about prescriptive guidance of one kind or another. Even the new developments described in Chapter 9, although based on fire safety engineering principles, are intended to produce solutions that can be picked up by the non-specialist designer with a minimum of adaptation to the context of a particular project.

In this chapter, the focus is on fire safety engineering methods for overcoming problems that the prescriptive guidance cannot easily solve. Various ways of addressing means of escape issues will be described. Some of the key data used in applying such methods will be given. The aim is to give the non-specialist reader a better idea of what is possible, and what it may involve. Obviously, the reader will not expect to become a fire safety engineer simply by reading this chapter, even though many of the calculations involved are simple in themselves.

10.1.1 The shortcomings of prescriptive guidance

The prescriptive guidance on means of escape design works by being conservative. It has to be conservative because it is meant to be applicable to a very wide range of situations, but not surprisingly, the result is that it can be very conservative sometimes in a particular case. One of the first benefits of a fire engineering approach is that the solution is tailored to a specific situation. This makes it possible to be less conservative.

Even knowing the circumstances of a particular building design there are many uncertainties. Much of the debate over the use of fire safety engineering reflects different views on the appropriate degree of conservatism in dealing with these uncertainties. There is also an issue about tailoring solutions closely to the design. Building uses change over time, and the fire strategy may have to be adapted to suit. Here the issue is whether the current manager/owner will recognise the need to adapt.

Construction industry interests see the conservatism of the prescriptive fire safety requirements as inefficiency, and would like to remove any measures that could be wasteful. The prescriptive approach has been based on past experience, and the principles that have been developed are not necessarily applicable to novel situations (it is not clear how well they address some common situations either). For most of the time and in most commonly found building situations, the harm caused by fire is within limits that society is prepared to tolerate. The regulatory system is responsible for maintaining a balance between the cost of fire precautions and the cost of fire. These costs are both the obvious direct costs of injury and loss, fire protection measures and the cost of running a fire service, and indirect or social costs that are reflected in investment decisions and public social and financial policy.

Every few years, this equilibrium between harm and preventive measures is seen to be lost. This is either the result of a fire or political pressure (often with a particular incident acting as a catalyst). Some aspects of the prescriptive rules are then altered. This process of adjustment can continue indefinitely, given the enormous number of variables and the changing expectations of society. Thus, while the prescriptions may be conservative, they may also be inadequate when faced with a shift in the way buildings are made or used.

The third drawback with the prescriptive system has been the inconsistency of approach to different aspects of the safety question. The prescriptive rules applied in the UK, and no doubt elsewhere, have simply accumulated. Being the ideas of different people at different times, responding to various political and technical concerns, they lack consistency, despite the best efforts of those concerned. One result is that the level of fire safety in some sorts of building is much higher than in others. For example, although the number of fire injuries in office buildings is very small in the UK, there is not much difference between the level of fire protection prescribed in larger offices and the level in, say, hotels or residential health care.

The inconsistency is also partly a result of the democratic process, and the intervention of political considerations. One objective of the regulatory system is to avoid certain kinds of multiple-fatality incidents that can cause public outcry and ministerial discomfort. This sometimes explains why prescribed measures can be illogical or inconsistent with others. They have been devised (perhaps in a hurry) to resolve one particular issue, at a time when that issue appears to be very important.

Although any system, prescriptive or performance based, can be affected by the political process, the issues could be more clearly set out in a mature performance based system. Comparison is easier if there are quantified goals as well as measures of performance. It may be that society wants office blocks to be x times safer than houses, but it would be nice to know.

Generally people do expect higher standards where the public or, to a lesser extent, any second or third party might be at risk. So the UK fire statistics indicate that one is much safer from fire in a hotel than at home, for example. This is tolerated partly because victims of dwelling fires are often seen as being the instrument of their own misfortune. It is more complicated than this, with factors such as the vulnerability of the exposed population, and their degree of control over the risk, being important.

10.1.2 *The cost of safety*

Fire safety measures:

- have a capital cost;
- may slow the construction process;
- require maintenance and repair;
- can reduce the functionality of the building;
- may never be required; *and*
- do not necessarily do much good if the need arises.

If no precautions were taken, fire losses would be much higher. But that does not mean that by spending much more than we do now, great savings would be made. The law of diminishing returns applies, and improvements to the system should be regarded as ways of achieving comparable levels of safety more economically.

10.1.3 *New buildings are different and need new approaches*

It has already been pointed out that the experience-based approach does not deal with novelty very well. Some insurance and fire service concern has been expressed that the prescriptive guidance has not 'kept up' with changes in the kind of buildings now built, and in the way they are constructed. The fear is that the guidance which developed out of experience with a building stock dating from the nineteenth century and the first half of the twentieth century does not address the characteristics of the buildings now being designed.

This is of relevance to fire engineering because the engineer often makes a comparison between the performance of his or her proposal and that of the prescribed equivalent (in the absence of explicit quantified objectives). The statistical evidence takes time to build up. So far, the UK statistics in the public domain do not support the contention that late twentieth century buildings are less safe. On the contrary, despite the large number of people potentially at risk in some of the very large 'new' sorts of building, the fire death rate has decreased.

But the property losses in fires have steadily increased. This is partly because the value of individual buildings can be much greater. The value is often not in the building fabric but in the goods or process it contains, and the consequences of business interruption. It is also true to say that, for some building owners, the return on investment in physical fire precautions is poor. For some companies the value of the buildings they occupy is a small part of their total assets.

After a significant rise in the post-war period, the fire fatality rate in private dwellings declined steadily in the 1990s. The annual death rate is in single figures in most other categories of building, and the law of diminishing returns applies to improving matters by means of constructional fire precautions.

However, the changes to what and how we now build are very obvious. Many more buildings contain very large numbers of people. Whole new classes of building have been created. New (or reinvented) building forms such as the atrium, and the large undivided space previously found only rarely, are commonplace. There has been a concentration of production capacity or storage capacity in some of these large buildings, which creates a new kind of financial exposure. Should the worst happen, a locality can lose employment and income to an extent that can have significant social repercussions.

10.2 A new approach to fire safety

Government policy-makers have been aware of the deficiencies of the old regulatory approach, and they have tried to address them by reforming Building Regulations in the 1980s. (The property protection issue has been deliberately set aside because the legislators' role has been seen as being in safety matters only.) As a result, throughout Great Britain it is now possible to depart from the prescriptive guidance on fire safety and propose a design based on a fire safety engineering strategy instead.

In several other countries the same desire to find a more efficient way of meeting society's building safety expectations (fire safety being an important element) has also caused major changes. The most radical has been in New Zealand, where the fire safety engineering approach appears to be almost as common as the code-based one. Parts of Australia are taking on a regulatory approach that recognises fire safety engineering. The code-writing bodies in the USA are all working on forms of new performance-based code, embracing fire safety engineering to varying degrees. Sweden now uses a performance-based code for building fire safety. Even the authorities in Hong Kong, where one of the most stringent prescriptive regimes has prevailed, are understood to be contemplating a new fire safety code of this sort.

10.3 Basic objectives

The approach taken in the UK by the prescriptive guidance for escape in case of fire, sets the agenda for a fire safety engineering approach. It is to ensure that building occupants can:

- get into a place of relative safety inside the building quickly; and
- from there have a route to the outside that is protected from the effects of fire, and which has sufficient capacity to serve the number of people who might use it, in a short time.

To this, one adds other measures, not within the scope of this book because they are not strictly about the escape process. They include:

- high levels of structural fire resistance;
- the sub-division of space into separate fire compartments;
- the specification of good reaction to fire properties for the surface of walls and ceilings;
- measures to avoid fire spread between buildings, mainly by radiation but also recognising the danger from flying embers;
- measures to avoid fire or smoke spread through large parts of a building via hidden voids, such as ceiling cavities; *and*
- the provision of fire service access to the building, and facilities within some larger buildings to assist fire-fighters.

10.4 What is fire safety engineering?

The following definition has been proposed in the International Standards Organisation's Technical Report ISO/TR 13387-1: 1999 Fire safety engineering – Part 1: Application of fire performance concepts to design objectives:

> '*Fire safety engineering* the application of engineering principles, rules and expert judgement based on scientific appreciation of the fire phenomena, of the effects of fire, and of the reaction and behaviour of people, in order to:
> – save life, protect property and preserve the environment and heritage;
> – quantify the hazards and risk of fire and its effects;

- evaluate analytically the optimum protective and preventative measures necessary to limit, within prescribed levels, the consequence of fire.'

The British Standard on fire safety engineering published as a Draft for Development in 1997, and subsequently revised and issued as a full standard BS 7974 in October 2001, contains the definition:

> '*Fire safety engineering* the application of scientific and engineering principles to the protection of people, property and the environment from fire.'

Fire safety engineering is a combination of several types of knowledge, with expert judgement. That judgement is both the individual engineer's, and the collective wisdom and experience of others, as embodied in 'rules' such as standards and other guidance.

In Webster's dictionary a definition of 'engineering' is:

> 'the application of science and mathematics by which the properties of matter and the sources of energy in nature are made useful to people.'

A key word here is 'useful'. Engineering is not an exact science, but it makes practical use of scientific knowledge. Like other engineers, the fire safety engineer operates in a complex world where problems are rarely expressed in clear-cut terms. The fire safety engineer has to make enough sense of it to be able to define, for example, the 'design fire' against which a protection system will be designed. This will obviously have to cover all situations that can be plausibly anticipated without being an unreasonable financial burden or design constraint.

In a similar way, a geotechnical engineer may be asked to design an 'earthquake resistant' structure without a precise definition of the earthquake that is to be resisted. There may be local codes prescribing forms of construction, or defining the accelerations and movements to be resisted, but the good engineer looks at specific aspects of site and structure too, just as the fire safety engineer does not simply apply the relevant code.

A good fire safety engineer will strive for a solution with Vitruvian 'firmness, commodity and delight'. The aesthetic impetus should come from the architect, but the fire safety engineer should rise to the challenge to help realise that vision.

10.4.1 *A code of practice for fire safety engineering*

A Draft for Development DD 240, on the application of fire safety engineering principles to buildings, was published by BSI in 1997. It was the outcome of several years' development. Since then the format and content have been reviewed. A Code of Practice BS 7974 on the application of fire safety engineering principles to the design of buildings, was published in 2001. The new standard will form the framework document linking a series of eight Published Documents (a particular category of British Standard Institution document). Each of these Published Documents, listed here, will deal with an aspect of fire safety engineering.

Published Document Part 0: Introduction: interaction of sub-systems, information bus, QDR, risk and probabilistic assessment (in brief)

PD0 provides a commentary on how the principles of engineering can be applied to fire safety, including an overview of the interaction of sub-systems, the selection of appropriate analytical approaches and the selection of methods.

Published Document Part 1: Initiation and development of fire within the enclosure of origin

This provides guidance on the evaluation of fire growth and or fire size in the enclosure where the fire starts, taking into account the four main stages of fire development:

- pre-flashover fire, including early growth and development;
- flashover;
- fully developed fire, in which all available fuel is burning;
- fire decay.

Published Document Part 2: Spread of smoke and toxic gases within and beyond the enclosure of origin

This provides guidance on the evaluation of:

- spread of smoke and toxic gases within and beyond the enclosure where the fire started;
- the characteristics of the smoke and toxic gases at a given point of interest.

Published Document Part 3: Spread of fire beyond the enclosure of fire origin

This provides guidance to permit the evaluation of:

- the severity of the fire in terms of temperature and heat flux within the enclosure;
- the ability of the elements forming the enclosure, directly or in part, to withstand exposure to the prevailing fire severity.

Published Document Part 4: Detection of fire and activation of fire protection systems

This provides guidance on the calculation of the following, with respect to time:

- detection of the fire;
- activation of the alarm and fire protection systems, e.g. sprinklers, smoke venting systems, roller shutters, lift controls, etc.;
- fire service notification.

Published Document Part 5: Fire service intervention

This gives guidance on the speed of build up of fire-fighting resources by the fire service, including the activities of in-house fire services, in respect of:

- the interval between the call to the fire service and the arrival of the pre-determined attendance;
- the interval between the arrival of the fire service and the initiation of their attack on the fire;
- the intervals relating to the build-up of additional fire service resources, if necessary;
- the extent of fire-fighting resources and extinguishing capability available at various times.

Published Document Part 6: Evacuation

This provides guidance on how to assess the evacuation time from any space inside a building. Once this has been established it can be compared with the outputs from sub-systems 1–4. Criteria for acceptance are given.

Published Document Part 7: Risk assessment, uncertainty and safety factors

This provides guidance on how to analyse the risk in a building and its contents, occupants and fire control systems, with the aim of determining:

- the frequency of occurrence of particular fire scenarios;
- the level of risk associated with fire and occupancy;
- extra measures required to reduce unacceptable risks.

The sub-systems idea

Parts 1–6 correspond to each of the sub-systems defined in DD 240. The sub-system idea is a way of handling the wide range of complex interactions that may need to be considered (Fig. 10.1). It is analogous to a set of computer

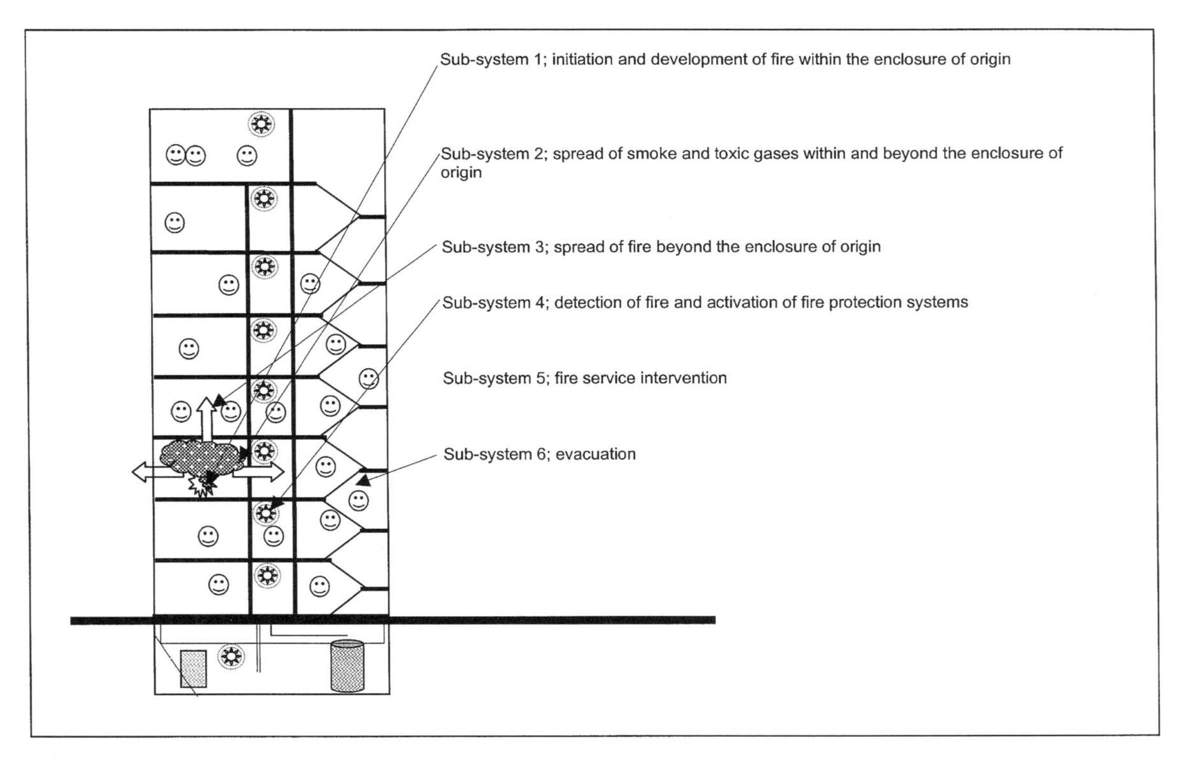

Fig. 10.1 Each fire engineering sub-system looks at a different aspect of the growing fire, or the response of people and systems to the fire threat.

processors, each working on its own part of the problem while taking input data from and sending outputs to, the others. For example, sub-system one is looking at the initiation and growth of the fire in the room of origin, while sub-system six follows the evacuation process. With each tick of the processor's clock, every sub-system takes another time step in following the course of events.

Part 0 explains the sub-systems concept and how they interact; it also provides a commentary on how the principles of engineering can be applied to fire safety, and discusses the selection of appropriate analytical methods.

The intention is that each of these parts gives advice on the technical ways and means for their part of the subject. Each of these sub-system Published Documents can be revised as and when the state of the art allows.

10.5 Fire safety engineers

10.5.1 Who is the fire safety engineer?

Fire safety engineering involves a wide range of subjects including:

- physics;
- chemistry;
- toxicology;
- psychology and behaviour sciences;
- mechanical, electrical, structural, and public health engineering;
- architecture;
- building construction and materials science;
- building legislation.

Until the 1990s, practitioners were likely to have started out from one of these areas. Now graduate fire safety engineers are entering the profession from universities in the UK, USA, Sweden, Australia and New Zealand.

In the USA fire protection engineers have been around for many years. Their function appears to have been slightly different from the British view of a fire safety engineer. Their role has mainly been in systems engineering, for example sprinkler systems and fire detection. They would not expect to be much involved with discussions with the architect about the layout and form of the building, as a fire safety engineer often is in the UK.

In the UK chartered engineer status under the Engineering Council regime has been available for suitably qualified fire safety engineering practitioners since 1999.

10.5.2 *What do fire safety engineers do?*

Fire safety engineers work in construction and other industries. Where buildings are concerned, fire safety engineers can be found working for:

Developers and building owners and/ or their architects	To develop fire safety strategy for a project and to assist in negotiating with the authorities for its acceptance
Regulators and enforcement bodies	To develop new approaches to regulations, or better standards or guidance; and to assess fire safety engineering proposals submitted to the authorities for approval
Construction product manufacturers	To define or assess fire safety performance levels, sometimes in the absence, or unsuitability of, standard test methods
Facilities management and operators of buildings	To assess fire risks and make proposals for physical or managerial measures to keep fire safety at an appropriate level

10.5.3 *The fire safety strategy report*

In the UK the typical fire safety engineering service, where a new building is being designed, is to develop a fire safety strategy. This is recorded in a report that is refined and expanded as the project progresses from outline to detailed design:

- it describes the fire safety issues and how they are to be addressed;
- it acts as a guide for the design team, by identifying standards or setting performance targets, e.g. for the capacity of a smoke extract system, and/or the fire resistance of elements of structure;
- it is the basis of the submission to the authorities for fire safety approvals.

The fire safety engineer takes part in design team discussions about broad issues, and will often support the architect in discussions with the authorities. The responsibility for obtaining approvals normally remains with the architect, because fire safety cannot be separated from other aspects of the design for which the architect has overall responsibility.

An engineering approach to developing an escape strategy starts by analysing the broad picture in a qualitative way to see what the fire safety issues are likely to be. The engineer may also have been briefed to look at other issues, such as property protection and business continuity in case of fire, but we will focus on the escape issue.

Involving the fire safety engineer at the scheme design stage of the project is usually thought to be a good thing. The fire safety engineer should be able to steer the architect away from building-in fundamental problems, and can explain the fire safety implications of options the architect will be considering, and the architectural implications of different fire strategies.

Involving a fire safety engineer only when a design is substantially complete, but has been rejected by the authorities, is not ideal. The engineer may be able to find a solution that is more economic than reverting to a code's prescription, but there is often little scope in these situations to avoid compromises that would have been unnecessary given earlier intervention.

The fire safety engineer is often asked to assess the 'approvals risk' associated with non-standard strategy options. Project managers weigh up the potential gains of a certain solution against the delay that might be involved in satisfying the authorities that it is acceptable.

10.5.4 The qualitative design review process

If the fire safety engineer has been part of the design team from the outline or scheme design stage, he or she will be involved in early reviews of the scheme. Initially this will involve the architect, and possibly the client's representative and some of the other consultants, depending on the type of building. It is a two-way process, with the fire safety engineer needing to understand how the building works at various levels, such as:

- the building's function, multiple use or mixed use, future trends;
- number and location of occupants, possible expansion;
- character of the organisation and the occupants, such as their familiarity with the building, their speed of response, and mobility;
- any factors that might complicate the evacuation process, such as process equipment to be shut down, or affiliations between people in different parts of the building that could lead to cross circulation in response to a fire alarm;
- circulation, e.g. restricted areas, daily patterns (and patterns which change at different times or seasons, etc.), emphasis on lifts or stairs;
- potential for fire service access;
- key dimensions that may trigger particular prescriptive requirements;
- location of significant fire hazards and fire loads;
- nature and degree of management control;

- form and any unusual aspect of the structure;
- environmental engineering, and particularly the ventilation strategy.

As well as gathering this information, the fire safety engineer will be reacting to it with comments on the fire safety implications. In the means of escape realm these might include:

- scope for phased evacuation and implications for management, in terms of controlling the process;
- the number and position of stairways;
- order-of-magnitude indications of exit route capacity;
- exit locations and the need for protected routes;
- interpreting existing guidance to define the needs of a code-compliant solution, and proposing outlines for fire safety engineered alternatives;
- assessments of the risk that non-code-compliant solutions will not be accepted by the authorities;
- scope for using ventilating system (or others) in a fire mode to improve means of escape.

This qualitative design review (QDR) process is described in the British Standard BS 7974 (the successor to DD 240, mentioned above). It should be appreciated that the process described in the BS is not one to be followed rigidly. Nor is it a once-only activity. It is more of a routine that will be repeated, at various levels of intensity, with every significant change in the design.

There is a checklist in BS 7974 of the topics that may have a bearing on the QDR (checklist for review of architectural design). The following list concentrates on the means of escape issues likely to be involved in a qualitative design review:

(1) character of occupancy:
   - what are the fire hazards likely to be?
     - ignition sources: industrial processes, cooking, certain types of plant such as generators, boilers and some high voltage plant, arson;
     - fuel load: areas where it is predictable, e.g. high rack storage, car parking, retail sales space, large circulation spaces, auditorium with fixed seating; other areas where it is predictable only within certain limits, e.g. offices, dwellings.
   - what are the occupants like?
     - familiarity with their surroundings: finding their way to exits, recognising fire alarm, knowing evacuation drill;

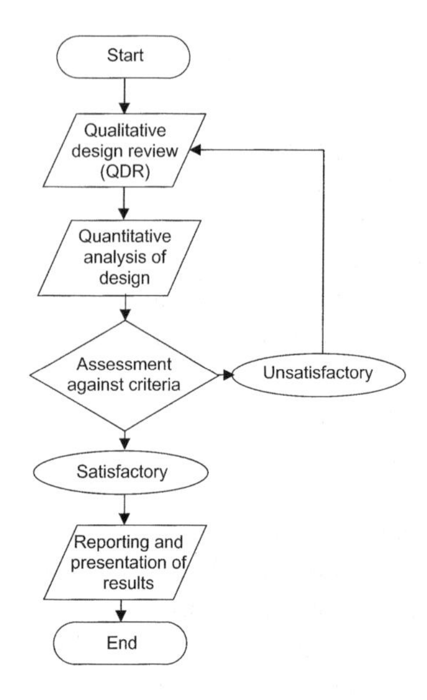

Fig. 10.2 Flowchart of the basic fire safety engineering process.

- capability: able-bodied, adult/child, bed-ridden, disabled in respect of mobility or vision or hearing or mental capacity; ratio of able-bodied to others; trained in fire procedure;
- awareness: sleep, attention focused on an activity such as dining or watching a show;
- social factors: disciplined hierarchy, e.g. hospital; affiliations, e.g. family groups; crowds or individuals.

(2) building form:
- size:
  - number of occupants;
  - length of escape routes;
  - potential complexity of wayfinding.
- height of storeys above or below ground level;
- potential for smoke spread:
  - large uninterrupted spaces; horizontally or vertically;
  - multicellular; spaces within spaces.
- 'natural' lines for sub-division into compartments.

(3) location:
- space (outside) for occupants to disperse;
- proximity of other buildings:
  - wind effects relevant to smoke ventilation;

   - emergency access to each façade;
   - potential fire sources near the building.
- urban/rural; availability of emergency services, water supply for fire suppression.

(4) functional issues:
- multi-purpose areas, mixed use areas;
- tenancies, and spaces subject to different rates of change, e.g. retail units in a shopping mall;
- circulation patterns: areas that are to be connected or separated.

The importance of a QDR is that, as a result of it, the principal members of the design team will have an understanding of the fire safety issues presented by this particular building. A responsibility that the fire engineer may have to take on (irrespective of the specifics of the professional engagement) is to make sure that team members are fully aware of the significance of issues arising from the QDR, in their special areas.

The QDR process is an ongoing one as the design develops, with the fire safety engineer taking the initiative to advise the team if developments require a significant change to previously agreed aspects of the strategy. The fire safety engineer will flesh out this understanding into a fire safety strategy that develops with the design.

A difficulty can arise if the fire engineer is engaged only to develop the strategy at the scheme design stage. If the design team does not follow that strategy in detailed design, or does not realise that they have not followed it, there can be trouble when the approving authorities draw attention to a discrepancy. It is therefore desirable to retain the fire safety engineer to review the design at the later stages, or better, to keep him or her involved in all design development.

10.5.5 Holistic or realistic design process?

It is often said that a fire safety engineering approach is one that considers every aspect of the fire safety problem. You may hear references to a 'fully fire-engineered' building. There is certainly a view that fire safety engineering is a matter of going to first principles, and designing or assessing every aspect of fire safety against quantified objectives. Some people believe that it is only acceptable to use fire safety engineering if it is used on everything.

The fully engineered approach is unlikely to be practical for a building project. It is normally too expensive, and may well be difficult for the authorities to accept if it is so unconventional that they cannot see how it compares with an ordinary solution.

There is a misgiving that mixing the prescriptive solution with a fire engineered one will leave some sort of dangerous gap. While this is possible, it is entirely avoidable if the engineer knows what he or she is doing. In 1999 the UK's Engineering Council recognised the fire safety engineering profession, and it became possible to be a chartered fire safety engineer. This was an important step in confirming standards of competence and addressing earlier criticism that the profession was unregulated.

The reality is that the fire safety engineer will build as much of the fire strategy around conventional prescribed solutions as is appropriate. If a prescribed solution from a code works as well as an engineered alternative, the former is likely to be adopted, for the simple reason that this reduces the risk that the project will not be accepted by the authorities. The engineered solution only comes into play where it is superior to the standard one in some crucial respect. The crucial aspect is usually a financial one, either to do with construction cost or with the value of the end result in terms of its utility or marketability.

The fire safety engineer should rely on the help of the design team to weigh this up. Which is cheaper when all the factors are taken into account (sometimes the risk of a delay represents a far greater cost than the capital saving)? Which best meets all the client's needs, taking whole life cycle costs such as staffing and maintenance into account?

The fire strategy will therefore normally present a combination of prescribed and engineered solutions, and it must spell out how these different elements fit together. In this sense it is 'holistic' and does not consider in isolation any elements that have a significant interaction.

10.6 'Time line' analysis of the escape process

Put at its simplest the matter of escape is one of getting people out of danger before the danger zone grows or spreads to engulf them.

(A) Time to get to safety < (B) Time for conditions to become dangerous

Having analysed a specific proposal, the fire safety engineer can tackle either side of the equation to see that the time to evacuate is reduced, or the onset of danger is delayed, as necessary.

Alternatively the fire safety engineer can compare (A) and (B) for the code-compliant solution to show that the performance of an equivalent engineered solution is similar (see Fig. 10.3 for an example of this type of comparison).

In Chapter 8 it was explained that the behaviours found during a fire are more complicated than the 'starting gun' model implied by standard guidance.

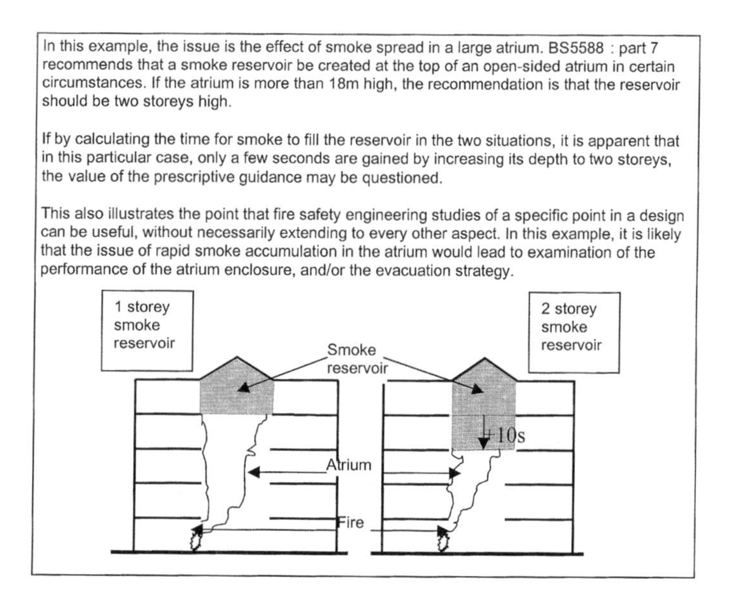

Fig. 10.3 Fire engineering analysis may show that code compliance is of little practical benefit in a given situation.

It has been known for some time that occupants do not respond to a fire alarm in a simple direct fashion. Arguably the only prescriptive British code that takes explicit account of this is BS 5588 : Part 7 on atrium buildings, in which the occupants' awareness and familiarity with surroundings help to determine which package of measures should apply. Other codes stipulate different maximum travel distances partly as a reflection of whether occupants are awake or asleep.

Otherwise the prescriptive model of escape used in Building Regulations guidance and elsewhere is very simple. No account is taken of the time for response, or any other pre-escape activity, other than by setting very short travel distances (which imply that most time is spent responding rather than moving). This should change when BS 9999 is eventually published (see Chapter 9).

10.6.1 The fire engineering time line evacuation model

BS DD 240 identifies distinct components of occupant behaviour that precede a committed movement to an exit route. There is a recognition phase and a response phase (Fig. 10.4).

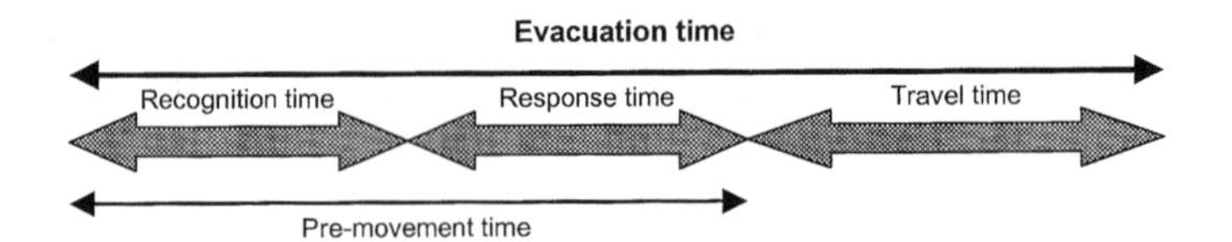

Fig. 10.4 Evacuation time line.

During the *recognition* phase the warning signs have to be:

- *Perceived* – hearing a noise/seeing a flame/smelling smoke;

and then

- *Interpreted* – deciding that the cue is not a telephone/a flickering lamp/burnt toast.

At the end of the recognition phase the occupant knows that there is a fire.

The *response* process is one of deciding what to do about it. It may involve investigation, asking other people what is happening or going to see first-hand. Then, before deciding to leave the area, the individual may gather up belongings, turn off equipment, finish a drink, or go and look for a friend or relative.

Travel time only begins when the decision is made to go to an exit, or other place of safety.

As noted above, an automatic fire detection system can have the effect of reducing recognition time, particularly for occupants not in the immediate vicinity of the fire. However the precise effect has as much to with human psychology as the mechanics of fire detection.

10.6.2 Design values for recognition time

DD 240 has suggested a range of recognition times for different occupancies. With the least effective type of alarm, a simple sounder, these vary from over four minutes in offices to over eight minutes in hospitals. This is a first attempt to codify these periods, and is certainly open to debate. It could be argued that a hospital would have a shorter time than an office, because of the training and motivation of staff. (When Part 6 of the series of published documents associated with BS 7974 is published, this information will be updated.)

DD 240 recommends a reduction in recognition time for a voice alarm, in the office case down to three minutes; and with a live voice message the office time could be less than a minute. DD 240 also advises on variations in recognition time for people who are near the fire and can see it, compared to occupants of other rooms who cannot.

In premises like shops, and cinemas where there are large public groups and a relatively small number of staff, the behaviour of the staff in an evacuation can have a very significant effect on all three phases of the process. In such situations, well-trained staff can give a clear lead to the public. Although there has been some research on this effect, there is no consensus on how to put a time value on it. It is difficult for the approving authorities to make an allowance for the intangible though positive effects of staff training, if they are being asked to accept a design that is dependent on a shorter than usual evacuation time. The draft proposals in BS 9999 might change this by suggesting some standards by which levels of building and staff management could be laid down.

10.6.3 Factors affecting the occupants' travel time to a place of relative safety

The prescriptive model

The travel time is obviously dependent on speed and distance. The regulatory model is based on a notional period of 2.5 to 3 minutes to reach a place of relative safety (not necessarily outside the building). Since 45 m is a commonly recommended travel distance limit, this could be taken to imply a speed of 18 m/minute or about 0.3 m/s. But this reverse engineering is misleading here. By setting very short travel distances the regulators were trying to make allowance for the evacuation time issues discussed above, even though the issues were not well understood at the time. The model was probably meant to take account of some (unspecified) pre-movement time, and of queuing time at the exit.

There are indications that the expectation was that the occupants would immediately make a concerted dash for exits. The concern may have been to avoid them waiting so long in a queue at exits that disorderly behaviour might start. Today the regulatory view is that occupants react quite slowly, and may be more likely to be overcome before reaching the exit than they are to be hurt in a crush. It is not possible to build an argument about speed or distance, on such an uncertain foundation. A new model for travel distance has been proposed in the draft of BS 9999 (see Chapter 9).

The second part of the prescriptive model concerns flow through openings, which can either be exit doorways or stairways. In general the advice tabulated in Approved Document B and the BS 5588 series of codes gives higher flow rates than accepted by equivalent codes used in some other countries. However, there is no record of stair capacity being inadequate in a code-compliant building in the UK.

The fire engineering model

There is good data on travel speed for able-bodied and disabled adults, moving on the level and up or down stairs. The relationship of travel speed to population density is also documented. In a confined space, such as a corridor, movement is halted when the density reaches about 0.25 m^2/person, and flow rate is greatest when density is about 0.5 m^2/person.

Where the population density is low, about 4 m^2/person or fewer than 2.5 people per 10 m^2, there is unlikely to be any significant queuing at exits. This is because people arrive at the exit(s) at a rate that does not exceed the flow capacity of the exit. This assumes that there is a practical minimum size of doorways. The estimation of travel time, at least on any one storey, is a simple matter of applying the average speed to the distance from the furthest point to the exit, or storey exit. A speed of 1 to 1.3 m/s can be used if the occupants are able-bodied.

However, other factors can effectively slow down average movement speed, even though the individuals are able-bodied. For example, if there are strong affiliations between individuals, such as family groups, they will try to keep together and will move more slowly.

There is a wide range of movement speeds among a population of disabled people. Some wheelchair users can move more quickly than able-bodied pedestrians, unless there are stairs to negotiate. A person using a walking frame, on the other hand, should be assumed to move at no more than half the speed of able-bodied occupants.

When there are more than about 2.5 people per 10 m^2 (an occupancy space factor of 4 m^2/person) the 'travel time' in a code-compliant layout depends on time spent queuing at the exit(s) rather than on travelling. Recognition of this fact should lead to a more flexible approach to determining travel distance in BS 9999, as discussed in Chapter 9.

Until that approach is codified, a fire engineering analysis can be used to show that travel distance can be extended.

Example

Space with design population of 100 people at a sufficiently high density for queuing time at the exits to be more significant than travel time to reach the exits.

Prescribed solution: two exits each of 900 mm, both able to cater for whole population (discounting rule).

Time for 100 people to pass through two such exits = 1.25 minutes

Clearly, if this order of queuing time is acceptable one could allow some occupants to travel for up to 1.25 minutes to reach the exit without

> increasing the overall time it would take them to leave the space. They will simply arrive at the back of the queue later than in the prescribed solution.
>
> Typical travel distance limits are 45 m. If these people can move at 1 m/s for 1:25 minutes (85 seconds) a distance of 85 m could be covered without affecting the time to clear the space.
>
> Under the normal discounting rule, only one exit might be available. Time for 100 people to pass through one such exit = 2.5 minutes. Logically it would be acceptable to increase travel distance up to the point where the stragglers take up to 2.5 minutes to get to the exit. According to this reasoning, even taking travel distance to 85 m allows evacuation of the space in half the time implied by the accepted discounting rule.

This is probably a longer travel distance than necessary, and longer than authorities would be inclined to accept, on such an argument. However it establishes a position from which a more 'comfortable' proposal might be agreed.

Stairway width

The estimation of stair width by hand calculation is not difficult, if the population on each floor is similar. But using the best data, e.g. from the *Society of Fire Protection Engineers' Handbook*, it tends to give wider stairs than the tabulated guidance in AD B.

AD B does not attempt to deal with big differences in population from one floor to the next. Where this is necessary it is possible to use computer models such as Simulex or Exodus to examine the effect of width on evacuation time. They will show how long it takes for each storey to clear into the stair(s). They are sensitive to the geometry as well as the width of storey exits and can therefore give useful guidance on layout to maximise flow.

However, the way that merging flows are treated by these models has a major influence on the pattern of escape. This is particularly important where people entering a stair meet others coming from another floor. Some models tend to give those in the stair priority, so that upper floors will tend to be clear before lower ones. Other models allow a 50/50 split, or prioritise those from the floor of fire origin, and this can delay upper floor occupants.

If it is important to estimate the time to clear a particular floor, these modelling effects should be borne in mind. It may be necessary to combine the simulation with some hand calculations. In reality one might expect people leaving the floor of fire origin to have the motivation to take priority where flows from different storeys merge. Furthermore those from the floor of origin are more likely to have started to move before most people on other storeys, and the merging problem may not affect the fire floor. This is especially so in open plan interiors where a fire or smoke is more visible, or in buildings with live voice alarm announcements; both factors reduce pre-movement time.

10.6.4 Other influences on time to evacuate

The times ascribed to each of these processes can be reduced or increased according to the features of the particular building. For example, automatic fire detection can reduce the recognition time, as can simple geometry which gives occupants a direct line of sight to the fire. The point has already been made that the way in which a fire warning is given also affects speed of reaction significantly.

Smoke control

Design calculations for smoke control are beyond the scope of this book. The main options, described in Chapter 8, are:

- physical containment;
- containment using pressure differentials;
- localisation – the 'cabin' concept;
- extraction – either mechanical or natural using channelling, ducts and/or reservoirs;
- dilution (rarely a practical option because of the high air volumes required).

Strategically smoke control systems can be regarded either as steady-state or transient in their effect. A steady-state system is designed to keep the smoke out of the occupants' way even when a fire has grown to fully developed stage. A transient system in effect delays the onset of untenable conditions for long enough to allow escape (with a safety margin) (see Fig. 10.5).

To be able to define the size of a fully developed fire the fire safety engineer has to know enough about the long-term nature and disposition of fire load to define a fuel-limited fire. This can be difficult, particularly if changes of use or layout in the future could upset the original assumptions. Alternatively a suppression system, such as a sprinkler system, is incorporated. This usually permits a maximum design fire to be specified. This can be justified because very rarely do sprinklered fires exceed a certain area of operation. The area is known from fire statistics for various occupancies and classes of sprinkler system. The potential peak heat release rate can be estimated per unit of floor area, and combined with the area of sprinkler operation to derive the fire size.

In the case of a transient system more complicated calculations are required, together with estimates of the rate of fire growth, to work out how long it may

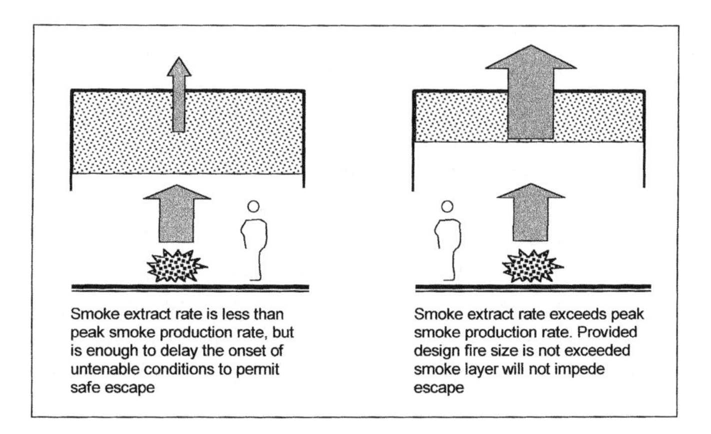

Fig. 10.5 Transient and steady-state smoke control.

take for, say, the layer of accumulating smoke to become deep enough below a ceiling to start to hinder occupants. There are quite widely accepted conventions for rates of fire growth, and accredited computer programs for working out the smoke fill time.

With these methods one must always be aware that the temperature of the smoke layer can be an impediment, even when the layer is some distance above head height. In fact the height of the layer is relatively unimportant when smoke temperature is assessed. If the average layer temperature is much more than 200°C travel for more than a very short while in the clear air below becomes uncomfortable.

This is not an issue where sprinklers are fitted because they will cool the plume of gases rising from the fire to 100°C or less. If they did not, one would see many more reports of large numbers of sprinklers operating, because each head is usually set to operate at around 70°C (Fig. 10.6). If gases spreading under the ceiling were at a higher temperature than this, all the sprinkler heads would open; but that only happens rarely when the system has failed to cope.

The prescriptive model assumes that the occupants are not exposed to smoke to any significant degree. While this may not reflect real experience in a significant number of fires, it does make it difficult to propose a scheme in which occupants may have to travel through smoke.

There is information about visibility in smoke, a characteristic measured by 'optical density'. Besides the attenuation of light, smoke irritates the eyes making them water and close up. So visibility tends to be reduced in every sense. There is also a rule of thumb that people are reluctant to go into a smoky space if they can see less than 10 m. This rule has a sound basis in practical research at the Fire Research Station. Another consideration ought to be whether it is possible to see the perimeter of the space from the starting point, so

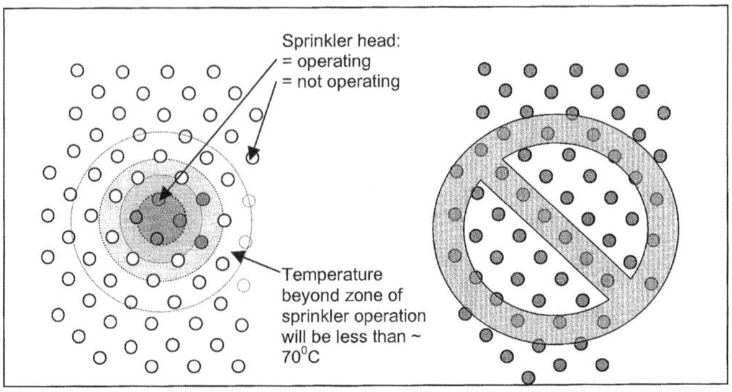

Fig. 10.6 Plan view of sprinkler array; the cooling effect of the spray means that only a few heads close to the fire operate. If large numbers of heads operate it is because the system has failed to control the fire.

that exits would be visible. Many spaces are much larger than 10 m or even 20 m across. The issue then becomes one of wayfinding in a thin smoky atmosphere.

Escape route lighting and wayfinding

Escape lighting and emergency lighting are terms with different meanings. Escape lighting is intended to allow occupants to leave an area where the ordinary lighting has stopped working. Emergency lighting is designed to allow occupants to continue to occupy, although they may not operate as efficiently as under normal lighting.

The need for either type of lighting may arise regardless of any fire safety consideration. It is often assumed that fires cause the lights to go out (they may do, but it is not common in the early stages of a fire when one expects evacuation to be taking place). There is also a misconception that escape lighting allows occupants to see in smoky conditions. Escape lighting design illumination levels are very low and are likely to have little effect on visibility if there is smoke.

Smoke tends to collect at high level, which is where light fittings are normally placed. If the lights were below the smoke layer they would be more effective. There is a new class of escape lighting system which uses this principle, similar to that of aircraft runway lighting, to guide occupants to exits. The lighting works, not by casting light over floor and wall surfaces to illuminate them, but by outlining the path to follow with bands or strings of light. The light sources in these wayfinding systems may be self-luminous materials, which emit light absorbed when normal lighting was available. There are also electric systems based on very low voltage sources such as light emitting diodes.

Audible wayfinding

Another wayfinding development, more recent than the illuminated variety, has been the use of sound sources to steer the occupant towards a fire exit. The system uses sounders with characteristic tones, such as a falling pitch or a rising one, to tell the listener in a stair that they should go downward, or upward. The intention is that the sounders can be electronically controlled using information on the location of the fire from an automatic detection system, to indicate which way to travel.

10.7 Footnote: architects and fire safety specialists

At present, fire safety engineering operates in the same context as the prescriptive fire codes. It is therefore often judged against those codes, even though the technical basis for the prescribed solutions may be less robust than the fire engineering principles. While that situation continues there will be problems of acceptance of fire engineered solutions that are matters of perception rather than of technical debate. This acts as a brake on the development of fire safety engineering. But it also forces the engineers to be more careful with the development of their proposals. This must be good for the long term improvement of the state of the art.

The development of fire safety specialists within the design team is another potential contributor to the de-skilling of architects. This applies to fully fledged chartered fire safety engineers as well as to other fire consultants. The best results are achieved where the specialist augments the architect's range of skills. If architects were simply to hand over responsibility for fire safety to others, the quality of design would be bound to be affected.

Fire safety has a major effect on the form of buildings. The architect should control that effect and balance it against all the other competing factors to get the best possible result. It may not be the best result in the fire safety engineer's eyes (although we naturally hope that it is), but it only has to be good. When the job is over, even fire engineers get a bigger kick out of the architecture than they do out of a sprinkler system.

Chapter 11
Management of Fire Safety

11.1 Introduction

Managing fire safety throughout the life cycle of a building is an essential element of an effectively engineered fire safety system. Specifically, with respect to life safety there are two key aspects where fire safety management has a role to play:

- to ensure that the fire safety measures that have been provided are kept in good working order;
- to initiate actions on the occurrence of a fire which will provide all the help and assistance that occupants need to reach a place of safety.

This chapter initially reviews the role of the fire safety manager and the ties between the design of fire safety systems and the management of such systems. A guide to conducting a fire risk assessment, a requirement of the Fire Precautions (Workplace) Regulations 1997 (as amended) for most places of work, is then presented. A fire risk assessment of a factory is illustrated in Appendix B.

11.2 Managing fire safety

11.2.1 Managing a facility

As the interaction between buildings, people and processes becomes more clearly evident, the effective management of facilities has become increasingly important. For large organisations, with extensive estate portfolios, this complex role is undertaken by facilities management departments, while for other organisations a traditional property management or maintenance management role may operate. Whatever the structure, generally the management of fire safety falls within the building or facilities management remit.

The importance of good fire safety management is slowly becoming recognised as a vital factor in reducing the numbers of, or effects, of fire. In many high-risk industries fire safety managers have for years had substantial powers and influence. But for some organisations the role of fire safety manager is relatively low in the management hierarchy, consequently with very little power. For most small organisations the role of fire safety manager is combined with other duties such as health and safety, or property security. This section provides an overview of the role of a typical fire safety manager.

11.2.2 The role of a typical fire safety manager

The role and responsibilities of a fire safety manager are delegated from the director or senior manager who is ultimately responsible for fire safety in the organisation. It is the responsibility of senior management to prepare a fire safety strategy tailored to the needs of the individual building and its occupants. An effective strategy should set out the fire safety management structure, practices and principles, arrangements for handling the recording of inspections, assessments, tests and evacuation drills, as well as the appropriate actions to be taken in the event of a fire. For some organisations such a strategy may be contained in a fire safety manual.

For simplicity, a fire safety strategy may be divided into four interlocking components as illustrated in Fig. 11.1:

- prevention;
- maintenance;
- training; and
- fire action plan.

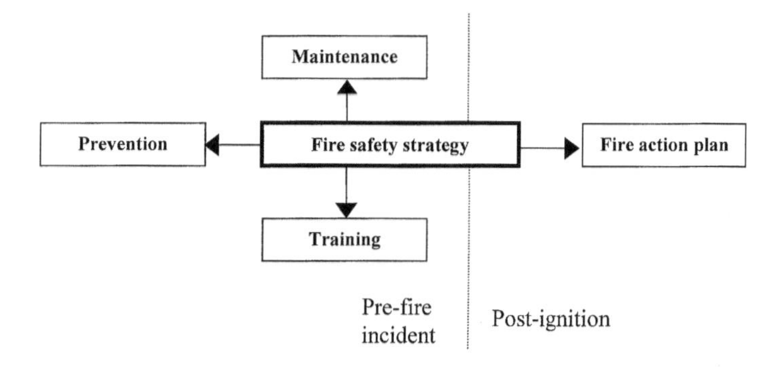

Fig. 11.1 Components of a fire safety strategy.

Typically the role of a fire safety manager consists of both routine 'day-to-day' responsibilities as well as tasks to be undertaken in the event of a fire. Here the duties of the fire safety manager are set out under the four sections of the fire safety strategy.

Prevention

The role of the fire safety manager starts with prevention. Effective prevention, together with maintenance, will mean that the likelihood of a fire occurring is reduced to the lowest practicable level. Prevention activities should include routine inspections to identify sources of ignition and potential ignitable items and the conduct of best practice housekeeping. Periodic risk assessments, risk assessment audits and regular liaison with the fire service also contribute to the prevention of fire.

A checklist of prevention tasks is given below. It should be the responsibility of the fire safety manager to carry out personally or manage the following tasks:

- Inspections
  - Identify sources of ignition (sources of heat which could get hot enough to ignite combustible items of fuel – see Table 11.1 for guidance).
  - Identify potential ignitable fuel items (items which will burn reasonably easily and are in sufficient quantity to provide fuel for a fire – see Table 11.1 for guidance).
- Housekeeping
  - Regular inspection of escape routes.
  - Ensure suitable security means are in place to reduce the chance of arson.
  - Regular removal of combustible waste.
  - Effective management of rubbish compactors and loading bay areas.
- Fire assessments
  - Conduct periodic fire risk assessments (see section 11.4).
  - Conduct an annual fire audit.
- Liaison with the fire service
  - Invite the crew from the local fire station to familiarise themselves with the layout of the workplace.
  - Inform the fire service of any temporary obstructions to fire-fighting operations.

Maintenance

Maintenance is concerned with the need to ensure that the fire safety measures and facilities in the workplace are in effective working order. It is likely that such

Table 11.1 Ignition sources and combustible materials.

Typical sources of ignition	Common ignitable fuel items
• Electrical equipment • Oil burning equipment • Open gas flames and gas burning equipment • Ovens, kilns, open hearths, furnaces or incinerators • Matches, lighters, candles, smoking materials • Flames or sparks from a work process • Sources of frictional heat • Ducts or flues • Light bulbs for fluorescent tubes if too close to flammable materials • Electrical extension cables • Faulty or damaged wiring or electrical equipment • Portable heater • Multi-point adapters in electrical sockets • Cooking equipment, including deep fat fryer	• Paper and card: books, files, waste bins, packaging • Fabric and clothing • Flammable liquid-based products • Flammable liquids and solvents • Flammable chemicals • Plastics, rubber and foams • Chipboard and hardboard
Reducing sources of ignition	**Minimising the risk of fuel ignition**
• Remove unnecessary sources of heat, or replace with safer alternatives • Remove naked flames and radiant heaters and replace with convector heaters or a central heating system • Ensure all electrical fuses and circuit breakers are of the correct rating • Operate a permit to work system for 'hot work' involving processes such as welding and flame cutting • Operate a no smoking policy except for designated areas • Make sure all smouldering material is properly extinguished • Take all necessary precautions to reduce the risk of arson	• Remove flammable materials and substances • Replace materials and substances with less flammable alternatives • Ensure that adequate separation distances are kept between flammable materials • Ensure flammable waste materials are stored and disposed of carefully • Remove furniture with damaged upholstery where the foam filling is exposed • Store highly flammable substances in fire-resisting stores

measures will be dependent on one another for their success, and regular checks and maintenance are necessary for them to remain effective. It is necessary for the fire safety manager to develop an inspection and service programme to manage this process. A checklist of maintenance tasks is shown below.

Weekly inspections should include:

- Fire detection and alarm systems
  - Check that fire detection and alarm system control panel is operating normally (test alarm system).

- Active fire fighting equipment and systems
  - Check that all fire extinguishers are in position and have not been discharged.
  - Check hose reels for correct installation.
- Facilities for fire-fighting
  - Inspect fire-fighting lifts, water supply hydrants, inlets and control valves.
  - Check that fire service access arrangements are free from obstructions and operable.
- Escape routes, signage and emergency lighting
  - Check that all exit and direction signs are correctly positioned.
  - Check that all emergency lighting systems are lit.
  - Check that all escape routes are clear of obstructions.
  - Check that all fastenings on escape route doors operate freely.
- Passive systems
  - Check that all self-closing devices and automatic door and shutter holders/releases work correctly.

The testing and servicing of equipment annually should include:

- Fire detection and alarm systems
  - Full service and test of system by a competent service engineer.
- Active fire-fighting equipment and systems
  - Full service and test of system by a competent service engineer.
- Emergency lighting
  - Full service and test of system by a competent service engineer.
  - Replace batteries in self-contained units.

Training

The fire safety manager is responsible for two levels of training: general staff training and training for individuals who have been designated by the fire safety manager to be responsible for fire safety in a specified area of the workplace. Training of staff is a legal requirement under the Fire Precautions Act 1971 and the Management of Health and Safety at Work Regulations 1999. At the very least, all staff should be made aware of details on the fire action notice. The training should include practical fire drill exercises to check people's understanding of the fire action plan. In addition, all new employees should be given suitable instructions on the location and use of escape routes and the locations and means of using the fire warning system on the first day of work, usually as part of an induction. It is normal for the fire safety manager to provide this training to the staff.

A checklist for staff fire safety training:

Staff should:

- be made aware of details on the fire action notice (see Table 11.2 for guidance);
- be given suitable instructions on the location and use of escape routes and the locations and means of using the fire warning system; *and*
- undertake regular practical fire drill exercises to check people's understanding of the fire action plan.

Table 11.2 Fire safety detail in a fire action notice.

What to do on discovering a fire
All employees should: • Know how to activate a break-glass manual fire alarm call point • Understand the existing and new (BS EN3) colour coding system for portable fire extinguishers • Know the different classes of fire • Know the types of fire extinguishers that may be used on the different classes of fire • Know how to use the fire extinguishers
What to do on hearing the fire alarm
All employees should: • Be able to recognise the sound of the fire alarm • Understand the importance of closing doors and windows as they evacuate the building • Know the location of all the escape routes in the building • Know the location of the assembly points
How to summon the fire brigade
All employees should understand: • That having dialed 999, they should request the fire service • They should then give the address of the premises slowly and clearly • Before replacing the receiver, they should wait for the fire brigade personnel to read back the address

Similarly, the training for designated individuals and other persons in a supervisory role is a statutory requirement under the Fire Precautions (Workplace) Regulations 1997 (as amended). The extent of the training depends on the size of the organisation and the nature of the employer's business, but the basic core of the training should include:

- detailed instructions on the evacuation procedure in the event of fire, including the roles and responsibilities of supervisors;
- hands-on training in the use of portable fire extinguishers and other fire-fighting equipment;

- an introduction to elementary fire science and fire safety legislation; *and*
- for some buildings, content retrieval training may also be undertaken.

Generally, such training is sourced externally, provided by the fire service or a fire safety training organisation.

The following represents a basic checklist for fire safety training of designated individuals:

Fire marshals should receive:
- detailed instruction on the evacuation procedure in the event of fire and their role and responsibilities;
- hands-on training in the use of portable fire extinguishers and other fire-fighting equipment;
- an introduction to elementary fire science and fire safety;
- buildings content retrieval training; *and*
- an overview of fire safety legislation.

There is also the issue of training for the fire safety managers themselves. Although there is currently no legal obligation for employers to provide any fire safety training for those appointed to be responsible for fire safety in the workplace, effective fire safety management cannot be expected without some element of training. The following represents a basic checklist for fire safety training of fire safety managers:

Fire safety managers should be given an understanding of:
- fire safety law and relevant British Standards;
- how to carry out risk assessments;
- the operation of fire safety systems;
- basic construction technology; *and*
- personnel management skills.

Fire action plan

The fire action plan should set out the responsibilities and actions that employees and other people in the workplace need to take in the event of a fire, to ensure that the workplace can be safely evacuated. If more than five people are employed the action plan must be in writing.

In a small workplace it may be little more than a list of instructions covering the points identified on a fire action notice (see Table 11.2). However, for large workplaces the plan will need to be more detailed, particularly for those buildings where the public are admitted, or for buildings occupied by people with disabilities.

For the fire safety manager the following tasks will be undertaken personally or will be delegated to a named person or persons:

- Fire-fighting:
  - First aid fire-fighting.
- Emergency management:
  - Calling the fire brigade and advising them;
  - Ensuring active systems have operated;
  - Ensuring that non-essential equipment is turned off;
  - Managing salvage team.
- People management:
  - Mustering occupants;
  - Organising evacuation;
  - Caring for displaced occupants.

11.3 Fire safety design and its effect on later management

The role of the fire safety manager is greatly influenced by the fire safety systems installed in the building during its construction or refurbishment. For a fire safety system to work effectively in the event of a fire it needs to have been both competently designed and implemented and also to have been properly managed. Communication between the fire safety engineer and the fire safety manager is essential if this is to be achieved.

11.3.1 The effect of design decisions on fire safety management

It is important that the fire safety engineer understands how the effectiveness of the design will influence fire safety management. To achieve this the designer needs to appreciate all the management implications in the design of the fire safety system. The implications need to be stated at the design stage and communicated to the fire safety manager. To aid this process the following needs to take place:

- Fire safety managers should be consulted at the earliest possible stage in the design process to ensure that the concerns of managers are taken into consideration. With a new building this may not be possible as such managers have yet to be appointed. In such cases representatives from the client need to be consulted. The use of operation and maintenance manuals is essential in the communication between designer, product supplier and the building management team.

- In order for the fire safety management team to carry out the jobs envisaged by the design, management must be given suitable resources and authority. Any restriction of resources is likely to prevent the effective management of the designed fire safety system and consequently the system will not operate as designed.
- All management activities associated with the management of the fire safety system must be clearly recorded for future managers, so that changes made which affect the operations of emergency systems are known and measures can be taken to correct the situation. Unrecorded changes may render an effective fire safety system non-effective.

11.3.2 What the designer expects of the fire safety manager

Conversely, the fire safety manager influences the effectiveness of the fire safety systems in the following ways:

- by ensuring that the fire safety systems are properly maintained and tested;
- by ensuring that the passive fire safety systems are not made ineffective, for example by blocking of escape routes or wedging open of fire doors;
- by ensuring that the building is used in the manner which conforms to the assumptions made at the design stage, for example the fuel load level is monitored, the level of occupancy is controlled and the types of occupancy are reviewed.

The degree to which fire safety management needs to be developed is dependent on the size and complexity of the building. The need can be established by an assessment of fire risk.

11.4 Fire risk assessment

The Fire Precautions (Workplace) Regulations 1997 (as amended) impose a statutory duty on the employer in charge of a workplace to undertake an assessment of the fire risk. This is the first time that the primary responsibility for fire safety has been placed upon the employer, while the fire authority acts as the enforcement body.

This situation has been created by the requirement for the UK, as a member state of the European Union, to implement the provisions of the 1989 Framework and Workplace Directives. The Directives introduced measures to encourage improvements in the health and safety of workers at work and prescribed the minimum health and safety requirements with which all workplaces (with limited exceptions) should comply.

The health and safety provisions of the Directives were implemented in the UK by the Management of Health and Safety at Work Regulations 1992 and the Workplace (Health, Safety and Welfare) Regulations 1992. Legislation to implement the general fire safety provisions of the Directives came into force on 1 December 1997 (The Fire Precautions (Workplace) Regulations 1997).

These Regulations, however, were considered by the European Commission to inadequately transpose the fire safety provision of the 1989 Framework and Workplace Directives. Specifically the commission noted that the Regulations:

- exempted many premises which should have been covered by the Directive;
- failed to reflect fully the unconditional nature of employers' responsibilities under the two Directives.

The amended Fire Precautions (Workplace) Regulations were laid before parliament on 7 July 1999 and came into force on 1 December 1999. The Regulations are divided into five parts:

- Part 1: Preliminary.
- Part 2: Fire precautions in the workplace:
  Regulation 3 – Application;
  Regulation 4 – Fire-fighting and fire detection;
  Regulation 5 – Emergency routes and exits;
  Regulation 6 – Maintenance.
- Part 3: Amendments of the Management of Health and Safety at Work: Regulations 1992.
- Part 4: Enforcement and offences.
- Part 5: Further consequential and miscellaneous provision.

11.4.1 Part 1

Part 1 includes the date of commencement (see above) of the Regulations and the definition of key terms. It is the interpretation of the term 'workplace' that is the most important. It is defined by the Regulations as 'any premises or part of a premise, not being domestic premises, used for the purpose of an employer's undertaking and which are made available to an employee of the employer as a place of work'. This is regardless of whether or not the employer is using the workplace as a profit generating business.

11.4.2 Part 2

Part 2 lists the excepted workplaces. These have been amended by Regulation 4 of the amending regulations and include the following:

- workplaces at construction sites;
- ships in the course of construction or repair;
- workplaces which form part of a mine facility;
- a workplace on or in an offshore installation;
- a workplace on or in a means of transport;
- any workplace which is in a field, wood or on other land forming part of an agricultural or forestry undertaking away from buildings.

In addition, the Regulations do not apply to:

- workplaces used only by the self-employed, *or*
- private dwellings.

Part 2 also implements measures in the Framework and Workplace Directives relating to fire-fighting and fire detection, emergency routes and exits and their maintenance.

Regulation 4: Fire-fighting and fire detection

A workplace must be equipped with appropriate fire-fighting equipment and with fire detectors and alarms. What is appropriate depends on:

- the size of the workplace;
- the use of the building housing the workplace;
- the equipment it contains;
- the physical and chemical properties of the substances likely to be present;
- the maximum number of people that may be present at any time (taking account also of persons other than employees).

Additionally, the employer must nominate employees to implement these measures and must ensure that the number of such employees, their training and the equipment available is adequate.

In addition it is required that:

- any non-automatic fire-fighting equipment should be easily accessible, simple to use and indicated by signs;
- necessary contacts with external emergency services, regarding rescue work and fire-fighting, are made.

In determining the appropriate level of fire precautions in individual cases, regard must be had to the results of the fire risk assessment.

Regulation 5: Emergency routes and exits

This regulation requires the employer to safeguard the safety of all employees in case of fire. To achieve this the following must be ensured:

- routes to emergency exits from a workplace and the exits themselves must be kept clear at all times;
- emergency routes and exits must lead as directly as possible to a place of safety;
- emergency doors must open in the direction of escape;
- sliding or revolving doors must not be used for exits specifically intended as emergency exits;
- emergency doors must not be locked or fastened in such a way that they cannot be easily and immediately opened by any person who may require to use them in an emergency;
- emergency routes and exits must be indicated by signs;
- emergency routes and exits requiring illumination must be provided with emergency lighting of adequate intensity in the case of failure of their normal lighting.

The number, distribution and dimensions of emergency routes and exits must be adequate having regard to the use, equipment and dimensions of the workplace and the maximum number of persons that may be present at any one time. Again the results of the fire risk assessment will determine the adequacy of the emergency routes and exits.

Regulation 6: Maintenance

The workplace and any equipment and devices provided in respect of the workplace under Regulations 4 and 5 must be:

- subject to a suitable system of maintenance;
- maintained in an efficient state;
- in efficient working order; *and*
- in good repair.

(Information on service, testing and maintenance can be found in section 11.2.2.)

11.4.3 Part 3

Part 3 amends relevant provisions of the Management of Health and Safety at Work Regulations 1992 in their application to the duties of employers in relation to general fire precautions. Regulation 3 of these Regulations requires the employer to make a suitable and sufficient assessment of the risk to health and safety to which his employees are exposed whilst at work. The effect of the amendment is that the health and safety risk assessment is expanded to now include an assessment of the risk to health and safety due to fire. If the employer employs five or more employees a written record of the risk assessment must be kept.

11.4.4 Part 4

Part 4 contains the enforcement and offence provision. The fire authority is responsible for enforcing the Regulations and is authorised to inspect the premises and examine the written fire risk assessment and the emergency evacuation plans.

The fire authority can issue an enforcement notice requiring works or improvements to be undertaken within a specified time, or a prohibition notice if the premises are considered potentially dangerous. Should the employer responsible for complying with the Regulations be found guilty, the penalties on summary conviction include a substantial fine and/or a prison sentence.

11.4.5 Part 5

Part 5 contains some consequential provisions, including the disapplication of section 9A of the Fire Precautions Act 1971 which required premises exempt from the need to have a fire certificate to provide means of escape in case of fire and means for fighting fire. Part 2 of the Fire Precaution (Workplace) Regulations 1997 (as amended) now applies to such exempt premises.

11.5 Conducting a fire risk assessment

Conducting fire risk assessments is an ongoing task of the fire safety manager which must be undertaken in an effective manner. An assessment of fire risk in the workplace should be nothing more than a careful examination of what could cause harm to people, so that the efficacy of the existing fire precautions can be assessed and extended if necessary, to prevent harm.

Assessments should not be over complicated. For small to medium-sized workplaces the fire risk assessment can be simple in its format, while for more complex workplace assessments the services of a fire safety engineer may need to be used.

Regardless of the complexity of the assessment it is suggested that it be structured into the following stages (see Fig. 11.2):

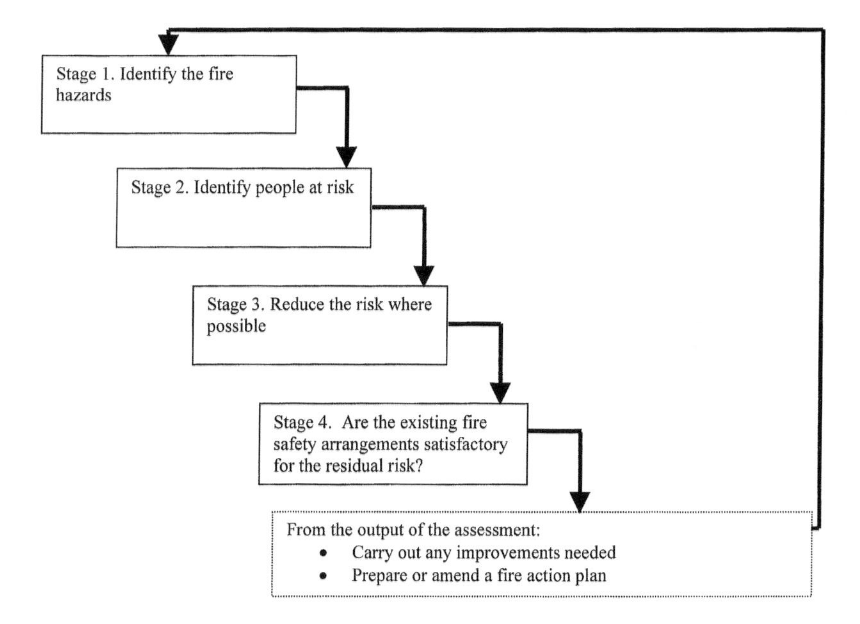

Fig. 11.2 Stages and cycle of a fire risk assessment.

The process of conducting the assessment is cyclical. It is wise to carry out a reassessment at least annually, or sooner if the nature of the work changes, there are major changes in work practices, staff responsibilities change or the workplace undergoes a refurbishment.

Before discussing the stages of a fire risk assessment in more detail it is important is clarify the two key terms: hazard and risk. A hazard is an object or situation with the potential to do harm. A hazard exists or it does not. Its existence is factual, not a matter of interpretation. A risk is the probability, or chance, or likelihood that a particular hazard will cause harm. A risk assessment is concerned firstly, with evaluating whether a hazard is significant and secondly, whether it is covered by satisfactory fire precautions so that the risk is small. For example, electricity can kill and so presents a hazard. In an office environment, however, provided that live components are insulated and metal casings properly earthed, the hazard is not significant and the risk is therefore small.

In conducting the assessment the following issues need to be taken into consideration.

11.5.1 Stage 1: Identify the hazards

This involves an inspection to identify both the sources of ignition present and potential ignitable items (see Table 11.1 above). All workplaces contain some hazards, but it is important to ignore the trivial ones and concentrate on significant hazards which could result in serious harm to people.

11.5.2 Stage 2: Identify people at risk

It is necessary to consider the risk to staff and other people who may be present. Certain people may be especially at risk from fire because of their specific role or workplace activity. Particular care needs to be given where:

- large numbers of the general public are present;
- sleeping accommodation is provided;
- people are unfamiliar with the building's layout and exit routes;
- people are working in areas where there is a specific risk from fire;
- people work in remote areas and have difficult or lengthy escape routes; *and*
- there are contractors who may be unaware of the fire risks in the workplace.

Additionally, it is possible that the population of the building may contain a number of people with physical and/or mental disabilities who may not react quickly to a fire emergency, such as:

- people whose hearing or eyesight is impaired;
- people suffering from heart ailments;
- pregnant women;
- people with learning disabilities or mental illness.

11.5.3 *Stage 3: Reduce the risk where possible*

For each of the identified fire hazards it is necessary to consider whether they can be removed, reduced, replaced or transferred, in the light of the risk to the identified people. Approaches to risk reduction are listed in Table 11.4.

11.5.4 *Stage 4: Are the existing fire safety arrangements satisfactory for the residual risk?*

The outcome of stage 3 will identify any residual risks which remain after the evaluation of approaches to reduce the risk. It then remains to evaluate whether the existing fire safety measures are adequate for the level of residual risk. An assessment of fire safety arrangements will cover the following issues:

- effective means for fire-fighting;
- fire alarm and detection systems;
- adequate means of escape;
- emergency lighting;
- evacuation signage;
- fire safety management procedures;
- levels of fire safety training given to staff.

The results from a fire risk assessment are then acted upon. Improvements to the fire safety measures to upgrade the fire safety arrangements to a level judged to be satisfactory for the residual fire risk should be undertaken. The fire action plan should also be based on the outcome of the fire risk assessment. For risk assessments conducted after the creation of the original fire action plan, it is important that the fire action plan is reviewed in the light of the latest fire assessment. The aim of the plan should be to ensure that in the event of a fire everyone in a building is familiar with the actions they should take to ensure that the workplace can be safely evacuated. In drawing up the plan the following factors observed in the assessment should be taken into account:

- the use to which the workplace is put;
- the means of raising the alarm;
- the number of people likely to be present and their location;
- people who are especially at risk and whether they are able to escape without assistance (special arrangements for people needing assistance);
- the level of training given to staff;
- the number of trained staff to assist members of the public in an evacuation;
- the presence of outside contractors.

For small workplaces a fire action plan may simply take the form of a fire action notice. In complex workplaces the fire action plan will be more detailed and may be compiled by a fire safety engineer.

The Fire Precautions (Workplace) Regulations 1997 (as amended) require that for employers employing five or more people, the fire risk assessment must be recorded. The record should indicate the significant findings of the assessment and any employees especially at risk. Three approaches to recording and evaluating fire risks are detailed in the next section.

11.5.5 *Possible approaches to recording and evaluating fire risks*

Recording the findings of a risk assessment can be undertaken in a number of ways. There is no one correct way and there are no national standards regarding fire risk assessment. In each case the appropriate methodology adopted should be practical, structured and effective for the workplace.

The 'simple' approach

The survey information can be recorded in a basic table as illustrated in Table 11.3. When using such a table, it may be appropriate to position the sources of ignition and combustible materials on a plan of the workplace. All decisions and judgements made in the assessment come from the common sense and knowledge base of the assessor. To aid the assessor in considering all relevant issues, an assessment checklist questionnaire can be used as shown in Table 11.4.

Table 11.3 Recording the findings – a simple table approach.

Significant hazards	Persons at risk	Existing measures to control hazards	Further actions required

Table 11.4 Fire risk assessment: checklist questionnaire.

Stage 1: Identify the fire hazards			
Combustible materials *(a 'no' answer highlights a concern)*			Notes:
Is the work activity free from the use of combustible materials?	Yes	No	
Is the workplace free from the accumulation of rubbish, waste paper or other materials which could be readily ignited?	Yes	No	
Are all flammable substances stored safely?	Yes	No	
Are the structure or fittings of the workplace free from excessive qualities of combustible materials?	Yes	No	
Is the workplace free from large amounts of stored combustible materials: timber, fabrics, furniture (containing foam padding) or furnishings?	Yes	No	
Is the workplace free from any other combustible materials which present a significant hazard?	Yes	No	
Sources of ignition *(a 'no' answer highlights a concern)*			Notes:
Are heating appliances fixed in position at a safe distance from any combustible material and fitted with suitable guards?	Yes	No	
Is the work activity free from the processes of incineration, cooking, welding, flame cutting, frictional heat or paint spraying?	Yes	No	
Is the workplace free from oil or gas burning equipment?	Yes	No	
Is all portable electrical equipment inspected regularly by a competent person?	Yes	No	
Is the wiring of the electrical equipment inspected regularly by a competent person?	Yes	No	
Are all flammable and combustible materials stored at a safe distance from light bulbs and fittings?	Yes	No	
Is the use of multi-point adapters and extension leads kept to a minimum?	Yes	No	
Are there suitable facilities for the disposal of smoking materials?	Yes	No	
Have measures been taken to reduce the risk of arson?	Yes	No	
Is there a no smoking policy throughout the workplace?	Yes	No	

Cont.

Table 11.4 Continued.

Stage 2: Identify people at risk			
People at risk *(a 'yes' answer highlights people at risk)*			Notes:
Are there staff who work in remote areas of the workplace, or in areas of high risk?	Yes	No	
Is there an area of the workplace which is used for sleeping purposes?	Yes	No	
Are there people in the workplace who are confined to bed or whose mobility is impaired?	Yes	No	
Is the workplace regularly used by people whose hearing or eyesight is impaired, or people suffering from heart ailments, pregnant women, or people with learning disabilities or mental illness?	Yes	No	
Do large numbers of people, particularly members of the public, occupy the workplace?	Yes	No	
Do people who are unaware of the fire risks and unfamiliar with the building layout and exit routes regularly use the workplace?	Yes	No	
Taking account of the identified people at risk evaluate the adequacy of the means of escape *(a 'no' answer highlights a concern)*			Notes:
Are there sufficient exits of a suitable width for the number of people present?	Yes	No	
Do the exits lead to a place of safety?	Yes	No	
Are all gangways and escape routes free from obstructions?	Yes	No	
Are all exit routes clearly lit?	Yes	No	
Are all escape route floor surfaces free from tripping and slipping hazards?	Yes	No	
Are all escape steps and stairs in good repair?	Yes	No	
Are all internal doors clearly labelled?	Yes	No	
Can all fire safety signs and fire exit notices be clearly seen?	Yes	No	
Do all exit doors open in the direction of travel?	Yes	No	
Are self-closing devices on fire doors in good working order?	Yes	No	
Can all doors used for means of escape purposes always be opened immediately without the use of a key?	Yes	No	
Is there always an adequate number of trained staff present to assist in an emergency?	Yes	No	

Cont.

Table 11.4 Continued.

Stage 3: Reduce the risk where possible *(a 'yes' answer highlights option for reducing risk)*			Notes:
Can any unnecessary ignition sources (sources of heat) be removed from the workplace?	Yes	No	
Can any combustible materials present be removed, or significantly reduced?	Yes	No	
Can the general housekeeping and the arrangements for the disposal of waste and rubbish be improved?	Yes	No	
Can additional measures be taken to prevent the occurrence of arson?	Yes	No	
Stage 4: Are the existing fire safety arrangements satisfactory for the residual risk?			
Fire safety measures *(a 'no' answer highlights a concern)*			Notes:
Are an adequate number of suitable fire extinguishers provided?	Yes	No	
Are the fire extinguishers and fire blankets suitably located and available for use?	Yes	No	
Does a competent person annually service the fire extinguishers?	Yes	No	
Where any form of fixed automatic fire suppression system is installed, is it in working order?	Yes	No	
Is the fire alarm system in working order?	Yes	No	
Can the alarm be raised without anyone being placed at risk from fire?	Yes	No	
Are the fire alarm call points unobstructed and clearly visible?	Yes	No	
Is the fire alarm system tested weekly?	Yes	No	
Where an automatic fire detection system is installed, is it in working order?	Yes	No	
Where escape lighting is installed, is it in working order and maintained regularly?	Yes	No	
Fire safety management *(a 'no' answer highlights a concern)*			Notes:
Are Fire Action notices clearly displayed throughout the workplace?	Yes	No	
Is suitable fire safety training given to employees?	Yes	No	
Are the duties and identity of employees who have specific responsibilities clearly understood?	Yes	No	
Are fire drills periodically conducted?	Yes	No	

The 'semi-quantitative assessment' approach

Another approach incorporates an element of semi-quantitative assessment. As can be seen in Tables 11.5 and 11.6, the findings are recorded as in the first approach, but in addition, an evaluation of the risk to people from the identified hazards is made by considering the combination of two factors. Firstly, the potential harm that may be caused by the hazard (the severity) and secondly, the likelihood of the hazard causing such harm. The severity and likelihood can be displayed as a matrix. The combination of the two evaluations generates a risk score which can then be translated to a low, normal or high risk.

Table 11.5 Risk matrix.

Likelihood / **Severity**	Very likely	Likely	Quite possible	Possible	Unlikely
Very severe	25	20	15	10	5
Severe	20	16	12	8	4
Moderate	15	12	9	6	3
Slight	10	8	6	4	2
Negligible	5	4	3	2	1

Risk category:	
(black)	High risk
(grey)	Normal risk
(white)	Low risk

This semi-quantitative assessment approach enables the assessment to be conducted using the four stage approach, but in addition, an element of risk evaluation is used. This approach is particularly useful in larger workplaces, where different areas may have quite different levels of risk and require individual risk assessments rather than one risk assessment for the whole workplace. Again, it may be appropriate to position the sources of ignition and combustible materials on a plan of the workplace, when conducting the assessment. The assessment checklist questionnaire may also be of benefit (Table 11.4).

The assessment worksheet approach

The third approach utilises a series of assessment worksheets. Each worksheet is completed in turn and the outcome identified on the summary sheet shown in Table 11.7. Worksheets 1 and 2 are concerned with the ignition sources and combustible materials present, as outlined in stage one of the fire risk assessment

Table 11.6 A semi-quantitative assessment approach.

<table>
<tr><th colspan="9">Fire Risk Assessment</th></tr>
<tr><td colspan="5">Assessor's name:</td><td colspan="2">Date:</td><td colspan="2">Assessment ref. no.:</td></tr>
<tr><td colspan="5">Assessment area/building:</td><td colspan="4">Persons at risk (affected groups)
A:
B:
C:</td></tr>
<tr><td>Identify hazard</td><td>Affected groups</td><td>Severity</td><td>Probability</td><td>RIsk level before controls</td><td>Control measures</td><td>Risk level after controls</td><td>Fire safety arrangements</td><td>Final risk level</td></tr>
<tr><td></td><td></td><td></td><td></td><td></td><td></td><td></td><td></td><td></td></tr>
<tr><td></td><td></td><td></td><td></td><td></td><td></td><td></td><td></td><td></td></tr>
<tr><td></td><td></td><td></td><td></td><td></td><td></td><td></td><td></td><td></td></tr>
<tr><td></td><td></td><td></td><td></td><td></td><td></td><td></td><td></td><td></td></tr>
<tr><td colspan="7"></td><td colspan="2">Page: of</td></tr>
</table>

Table 11.7 Worksheet assessment approach.

Fire Risk Assessment Summary Sheet	
Worksheets	Outcomes
1. Ignition sources	Satisfactory Unsatisfactory
2. Combustible materials	Satisfactory Unsatisfactory
3. Risk to life	High Normal Low
4. Means of escape	Above norm Norm Below norm
5. Passive protection	Above norm Norm Below norm
6. Active fire-fighting measures	Above norm Norm Below norm
7. Detection and communication	Above norm Norm Below norm
8. Fire brigade facilities	Above norm Norm Below norm
9. Housekeeping	Above norm Norm Below norm
10. Prevention, management and training	Above norm Norm Below norm

Definition of outcomes:
Ignition sources
Unsatisfactory – can they be eliminated, controlled, avoided or removed?
Combustible materials
Unsatisfactory – can they be controlled or removed?

Above norm: Statutory defined standard exceeded
Norm: Statutory defined standard met
Below norm: Statutory defined standard not achieved – upgrade required

structure. Worksheet 3 assesses the risk to people (stage two). Worksheets 4 to 10 evaluate the existing fire safety arrangements (stage four). These assessments are made against norms derived from statutory guidance. From the assessment output, methods of reducing fire hazards, upgrading fire safety arrangements and the effects on life risk are evaluated. This assessment approach is required to be conducted by a competent person with a good technical knowledge of fire safety and an understanding of fire safety legislation. Fire safety engineers and fire service assessors are currently using such assessment schemes.

11.6 Conclusion

Thus, the Fire Precautions (Workplace) Regulations 1997 (as amended) impose a statutory duty on the employer in charge of a workplace to undertake an assessment of the fire risk. Where five or more people are employed, the risk assessment must be in writing and the fire authority is authorised to inspect the premises and examine the written fire risk assessment and the fire action plan. Failure to comply with this statutory duty can result in the issue of an enforcement notice by the fire authority requiring works or improvements to be undertaken within a specified time, or the issue of a prohibition notice if the premises are considered potentially dangerous. Additionally, where court action is taken against an offending employer the penalties on summary conviction include a substantial fine and/or a prison sentence.

Therefore, it is especially important that employers in small and medium-sized businesses have a clear understanding of the law in the area of workplace fire safety. In many cases it should be possible for managers in small businesses to carry out their own fire risk assessments. In larger enterprises, knowledge of the law and the assessment procedures will provide managers with greater awareness of the work which is being undertaken by specialists on their behalf.

Appendix B contains a worked example of a fire risk assessment of a small factory unit. The purpose of this case study is to illustrate one practical approach to conducting a fire risk assessment and to provide the tools whereby the management of a small business might carry out their own assessment. The checklist and proformas used in this case study may be used as shown or may be adapted to suit an individual's own workplace.

Appendix A

Means of Escape Case Study

This case study looks at the design of the means of escape in a simple five-storey office block of rectangular plan (see Fig. A1). The block, which has external dimensions of 44 m × 17 m, consists of a basement, ground floor and four upper floors and is to be occupied by a single company. The ground floor is to be used also for corporate entertaining.

Step			
1		***What is/are the main use(s) to which the building will be put?***	PG 3 Office
	Assessment of use (Purpose Group)	Are there any ancillary uses? Is the building divided into different occupancies/uses? Is the building divided into different compartments with different uses?	No No No
2	Calculation of occupant capacity	***Estimate the total number of people who are likely to be in the building***	
		Are the numbers known accurately (such as in a theatre or restaurant)? Are floor space factors available from actual data taken from similar premises? If not, use floor space factors from Table 1 of AD B1.	See schedule of accommodation below. No See schedule of accommodation below.
		Calculate number of occupants per floor/use/occupancy/compartment	See schedule of accommodation below.

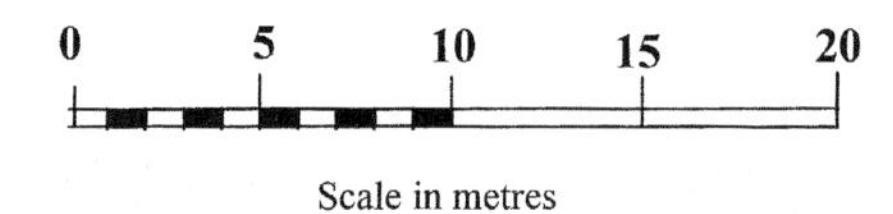

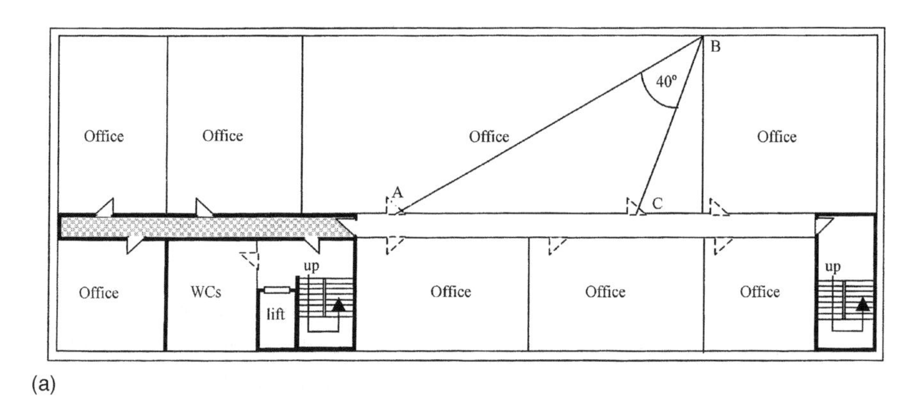

(a)

Floors 1 to 4

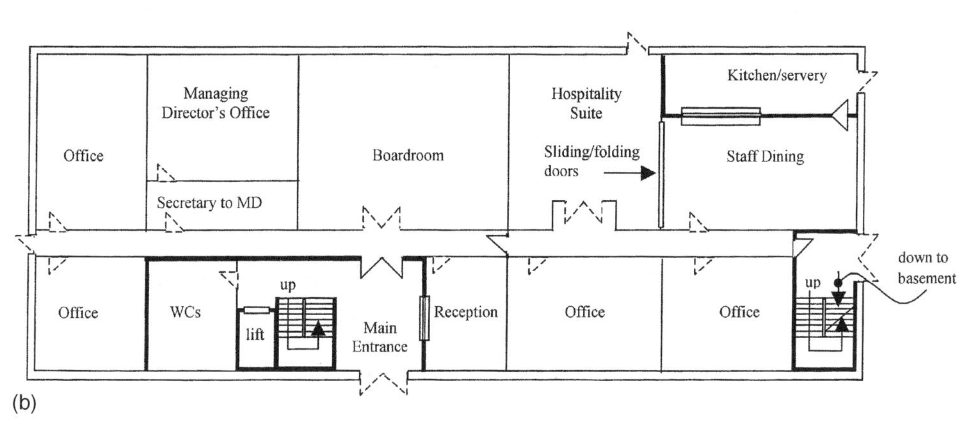

(b)

Ground Floor

Note: Fire doors indicated by solid lines and fire resisting construction by bold lines.

Fig. A1 Case study office block – ground and upper floors.

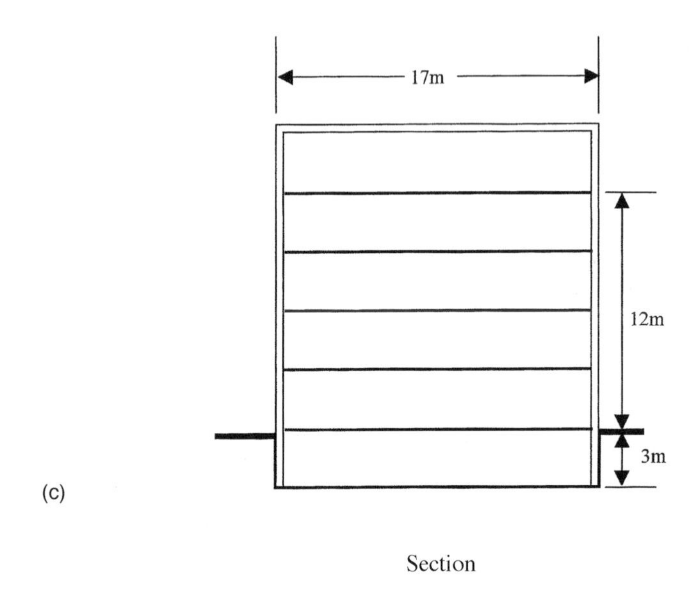

(c)

Section

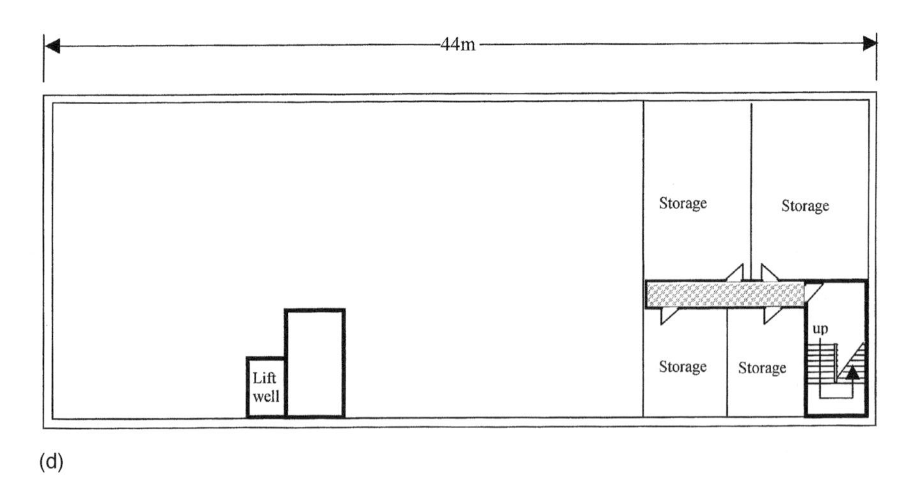

(d)

Basement

Fig. A1 Case study office block – Basement and section.

Schedule of accommodation/occupancy calculation

(a)	(b)	(c)	(d)	(e)	
Floor	**Room**	**Area**	**Floor space factor (from Table 4.1) m^2 per person**	**Occupancy = (c)/(d)**	**Comments**
Ground	General offices (4 no.)	$6 \times 9 + 6 \times 19 =$ $168\,m^2$	6.0	28	
Ground	Office – MD			1	Design occupancy
Ground	Office – secretary to MD			1	Design occupancy
Ground	Reception			3	Design occupancy
Ground	Staff dining	$6 \times 10 = 60\,m^2$	1.0	60	Seating area (designed to be used independently and can be linked to hospitality suite)
Ground	Kitchen			4	Design occupancy
Ground	Hospitality suite	$9 \times 8 - 1 \times 2 =$ $70\,m^2$	1.0	70	(Designed to be used independently and can be linked to staff dining)
Ground	Boardroom			20	Design occupancy (20 board members)
			Total ground floor	187	
First	Office (8 no.)	$9 \times 42.5 + 5.5 \times 30 =$ $547.5m^2$	6.0	92	General office occupancy based on floor space factors
Second	Office (8 no.)	$9 \times 42.5 + 5.5 \times 30 =$ $547.5\,m^2$	6.0	92	General office occupancy based on floor space factors
Third	Office (8 no.)	$9 \times 42.5 + 5.5 \times 30 =$ $547.5\,m^2$	6.0	92	General office occupancy based on floor space factors
Fourth	Office (8 no.)	$9 \times 42.5 + 5.5 \times 30 =$ $547.5\,m^2$	6.0	92	General office occupancy based on floor space factors
Basement	Storage			0	Only occasional visits to archives. No permanent staff presence
			Total all other floors	368	

Horizontal escape

Step			
		Check maximum travel distances from scaled floorplans	
3	Assessment of escape routes and travel distances	Are single and multi-direction paths of travel within the distance limitations given in AD B1 Table 3?	See Table 5.2. PG 3 – Office use Maximum travel distance (m) in: One direction 18 – Multi-direction 45
		Are assumed alternative paths of travel true alternatives?	Check '45°' rule. Floors to 4: Angle ABC = 40° Distance BC = 10 m (i.e. $\leq$ 18 m) therefore OK.
		Adjust floor layout and position of storey/room exits to accommodate maximum allowable travel distances.	No adjustment necessary since all paths of travel are within specified distances.
		Check that assumed numbers and widths of exits can cater for expected occupant numbers at each level	
4	Assessment of numbers and widths of storey/room exits	Are enough exits provided from the room/storey? (See AD B1 Table 4).	up to 60 persons – 1 exit (all rooms comply) 61 to 600 persons – 2 exits (all floors comply + hospitality suite)
		Are widths of exits sufficient to take the expected numbers of occupants (including the possibility of discounting)? (See AD B1 Table 5).	Doors (see Table 5.3): • to offices, WCs and kitchen: 750 mm min. • across corridors and onto staircase enclosures floors 1 to 4: 850 mm min. • external, ground floor: 850 mm min. • others are double and exceed these figures. Exit corridors (see Table 5.3): • min. 1200 mm to allow for disabled access (this also works for discounting).
		Adjust floor layout and numbers/widths of storey/room exits to accommodate expected numbers of occupants.	No adjustment necessary.
		Establish if any of the identified horizontal escape routes need to be fire protected and/or separated from other parts of the building	
5	Access need for protection and/or separation of escape routes	Do any dead-end situations exist?	Yes. Floors 1 to 4 (dead-end exceeds 4.5 m in length). Provide fire-resisting construction to separate these areas from the remainder of the exit route. Basement (see below).
		Are there any corridors serving bedrooms?	No
		Are any corridors common to two or more different occupancies?	No
		Do any corridors connect alternative escape routes?	Yes. Provide sub-division door roughly at mid-point of corridor connecting alternative exits.
		Are any basements present?	Yes. Provide protected corridor between stairs and basement accommodation.
		Are any inner rooms present?	Yes. MD's office. Either provide $0.1\,m^2$ vision panel in door or wall of MD's office or fit suitable automatic fire detection and alarm system.

Vertical escape

6	Escape stair design – numbers of stairs	***Assess the number of escape stairs needed***	
		Are sufficient storey exits provided? (See steps 3 and 4)	Yes
		Is a single stair acceptable?	Only from the basement (all parts of the basement accommodation are within 18 m of the protected stairway). Upper floors need minimum two stairways.
		Are independent stairs required (mixed occupancy buildings)?	No
		Are fire fighting stairs needed?	No
		Will it be necessary to discount any of the stairs?	Yes (stairs are not lobbied, see below for influence on stair width.)
		Is access needed to basements?	Yes. Only one of the two stairs connects to the basement.
7	Escape stair design – widths of stairs	***Assess widths of stairs needed and check that assumed numbers/widths can cater for expected occupant numbers***	
		Check that minimum stair widths are provided. (See AD B1 Table 6).	Discounting of one stair is necessary. Therefore each stair independently must be capable of taking the full building occupancy. Using Table 5.5 and schedule of accommodation above: Max. numbers using stair = 375 (sum of floors 1 to 4) but this will only occur on the flight from floor 1 to ground. (The width of the stairs could be reduced as it rises but this is not normally cost effective since it affects the standardisation of the design.) Occupant numbers >220. Therefore use AD B1 Table 7 or Table 8 (see below).
		Is escape based on simultaneous evacuation? (See AD B1 Table 7).	From Fig. 5.5 maximum numbers on 1000 mm stair = 270. Therefore use formula (5.1) to calculate minimum stair width. $P = 200w + 50(w - 0.3)(n - 1)$ where P = the number of people that can be served by the stair, w = the width of the stair in metres, *and* n = the number of storeys in the building. $368 = 200w + 50(w - 0.3)(4 - 1)$. Therefore $w = 413/350 = 1180$ mm (say 1200 mm)
		Is escape based on phased evacuation? (See AD B1 Table 8)	This is an alternative to the above but would require suitable management procedures to be put in place. Additionally, stairs would need to be separated from the accommodation at each floor level (except the top storey) by a protected lobby or corridor. This is already the case at the west end of the building but would require the addition of protected lobbies at the east end. There are also a number of requirements concerning fire detection and alarm, etc. that would need to be met. Using Fig. 5.6 the minimum stair width can be 1000 mm.

Cont.

		Assess the need for protection to the escape stairs	
		Is an unprotected stairway acceptable?	No.
		Is the stairway a protected shaft or a fire fighting shaft?	A protected shaft but not a fire fighting shaft.
8	Escape stair design – protection of stairs	Are access lobbies or corridors needed to protect the stairway?	Not unless phased evacuation is called for in the design
		Is the stairway to be used for anything other than access and escape?	Yes. WCs are entered off the stairway (but will not be used as cloakrooms). At ground floor level the main entrance lobby will not be used for office purposes (there is a separate reception office).
		Are the external walls of the stairway protected from fire occurring within the building?	Yes. The stairways are flush with the external walls and there are no projecting walls or re-entrant angles.

Exits from escape routes

		Check design of exits from the building/protected stairway	
		Does the protected stairway discharge directly to a final exit?	Yes
		Does it discharge by way of a protected exit passageway to a final exit?	No
		Is the position of the final exit clearly apparent to users?	Yes
9	Final exits	Is width of the final exit adequate for the expected occupant numbers?	Some assumptions need to be made regarding the number of people that will use each of the final exits shown on the plans. Taking the worst case scenario (assuming that kitchen staff will use their own escape door) and assuming that the main stairway is not accessible: Floors 1 to 4, east end stairway = 368 Hospitality suite, (say half of occupants use east end) = 35 Staff dining = 60 2 no. offices, east end = 14 Total = 477 Therefore, minimum width of exit from east end stairway to external air = 477 × 5 mm = 2385 mm A similar calculation can be done for the main stairway assuming that the east end stairways is not accessible: Floors 1 to 4, main stairway = 368 Hospitality suite, (say half of occupants use main exit) = 35 (MD and secretary and west end offices will use west end exit) Staff dining = 60 Reception = 3 Boardroom = 20 Total = 486 Therefore, minimum width of exit from main stairway to external air = 486 × 5 mm = 2430 mm
		Is final exit sited clear of any risk areas so as to ensure safe and rapid dispersal of occupants?	Yes

Cont.

10	External escape routes, lighting and signage	***Assess the design of any external escape routes, the provision of lighting, and escape route signage***	
		Are there any external escape routes?	No. All final exit doors lead to open space away from building.
		Do they pass over any flat roofs?	No.
		Are all escape routes adequately lit?	Adequate artificial lighting is provided on all parts of horizontal and vertical escape routes. Emergency lighting is provided (see Table 5.6) | as follows: • Basement storerooms and escape corridor, • Boardroom, hospitality suite, staff dining and all offices exceeding 60 m^2 floor area • All internal corridors • Toilet accommodation
		Are all escape routes distinctively and conspicuously marked by emergency exit signs?	Yes. See BS 5499 Fire safety signs, notices and graphic symbols, Part 1: 1990 Specification for fire safety signs.

Completion of the 10 stages outlined above will ensure that the means of escape meets the minimum guidance in Approved Document B1 for this particular building regarding the size and siting of the various elements (horizontal and vertical escape components and final exits). There are of course many more decisions to be taken regarding the general constructional provisions that apply to all buildings irrespective of their use, such as:

- The standard of protection necessary for the elements enclosing the means of escape.
- The provision of doors.
- The construction of escape stairs.
- Mechanical services including lift installations.
- Protected circuits for the operation of equipment in the event of fire.
- Refuse chutes and storage.
- The provision of fire safety signs.

Details of these provisions are given in the text above. An example of a door schedule is provided below which differentiates between fire resistant and non-fire resistant doors in the case study building (see Table 5.8).

Floor	Door no.	Description/location	Fire resistance (mins)	Remarks
Ground	GE1	Main entrance	N/A	Glazed double doors
	GE2 to GE4	Doors to final exits	N/A	Glazed single leaf doors
	GO1 to GO5	Office doors (not including MD's office)	N/A	Flush doors
	MD1	Door to MD's office	N/A	Flush door with 0.1 m^2 vision panel
	R1	Reception door	N/A	Flush door
	R2	Service hatch	FD30S	Fire shutter held open by automatic release mechanism actuated by the fire detection and alarm system
	B1	Boardroom doors	N/A	Double flush doors
	HS1	Hospitality suite doors	N/A	Double flush doors, outward opening
	SD1	Staff dining doors	N/A	Flush door
	K1	Kitchen/servery door	FD30*	Double swing fire door with vision panel*
	GWC1	Door to WCs	N/A	Flush door with self-closer
	GFD1	Door across corridor	FD20S	Self-closing fire door with cold smoke seals and vision panel, sub-dividing corridor which connects alternative exits
	GFD2	Door to east end protected stairway	FD30S	Self-closing fire door with cold smoke seals protecting entry into protected shaft.
	GFD3	Doors to main protected stairway	FD30S	Double self-closing fire doors with cold smoke seals protecting entry into protected shaft.
1 to 4	UWC1	Door to WCs	N/A	Flush door with self-closer
	UFD1	Door across corridor at dead-end	FD20S	Self-closing fire door with cold smoke seals and vision panel, sub-dividing dead-end portion of corridor from remainder
	UFD2	Door to east end protected stairway	FD30S	Self-closing fire door with cold smoke seals protecting entry into protected shaft
	UFD3	Door to main protected stairway	FD30S	Self-closing fire door with cold smoke seals protecting entry into protected shaft
	UFD4 to 6	Doors to offices in dead-end	FD20S	Self-closing fire doors with cold smoke seals protecting dead-end corridor
	UO1 to 6	Office doors	N/A	Flush doors
Basement	BFD1	Door to east end protected stairway	FD30S	Self-closing fire door with cold smoke seals protecting entry into protected shaft
	BFD2 to 4	Doors to storerooms in basement	FD20S	Self-closing fire door with cold smoke seals protecting basement corridor

* *See case study conclusions below*

Case study – conclusions

This case study is not intended to be a definitive solution to means of escape in the subject building. It is merely an example of the application of the rules governing means of escape to be found in Approved Document B1.

A close analysis of the design decisions taken will reveal a number of areas where the building control body may require further information/clarification:

- The client is to be the eventual user of the building. Therefore he will have certain information regarding the use of some of the rooms (e.g. the boardroom is large enough (99 m^2) to house many more than the design occupancy of 20) and the implications this will have for the total building occupancy.
- The basement rooms are designated as 'storage' because the client's business involves the possession of large quantities of archive material. The building control body may need assurances that they will not be used as offices since the means of escape design ignores the presence of people permanently in the basement.
- The main entrance lobby is larger than 10 m^2 so it must not be used as a reception area or waiting room. Visitors would need to be shown to another room for this purpose.
- The construction surrounding the kitchen in the staff dining room does not need to be fire resistant according to Approved Document B. However, BS 5588 : Part 11 recommends in Table 11 'robust construction having a fire resistance of 30 minutes' for 'Kitchens (separately or in conjunction with an associated staff restaurant or canteen)'. Furthermore 'Any openings in the required construction should be protected by doors having a similar standard of resistance'. This is further reinforced in Table 9 from BS 5588 : Part 11 where the minimum standard for fire doors in such enclosures is given as FD30. This implies that the servery needs to have an automatic fire shutter in a similar manner to that described for the reception service hatch.

The general comment that the fire authority can require standards over and above those needed for Building Regulation compliance has been made in earlier chapters of this book. It should not be a problem, provided that adequate consultation has taken place between all interested parties before the design has progressed to such a stage that changes become prohibitively expensive and/or time-consuming.

Appendix B
Fire Risk Assessment Case Study

The purpose of this case study is to illustrate one practical approach to conducting a fire risk assessment. The checklist and proformas used in this case study may be used as shown or may be adapted to suit an individual's own workplace.

The assessment uses the first of the approaches described in section 11.5.5 above and is suitable for all small to medium-sized workplaces where the arrangement of the premises are not complex. The fire risk survey is initially undertaken using the fire risk assessment questionnaire as a guide. The sources of ignition present and potential ignitable items are recorded on floor plans of the premises. From the information gathered the risks are evaluated and the outcomes of the assessment are presented in a summary table. The 'actions required', listed in the summary table, represent the measures that need to be taken to upgrade the fire safety arrangements to a level judged to be satisfactory for the residual fire risk. The 'actions required' are prioritised and listed on the first page of the assessment.

The workplace in this case study is a coach building firm which customises lorries and vans. The factory unit is 800 m^2 consisting of office accommodation, workshops and a storeroom. The factory has a portal steel frame structure with full height brick and block cladding. The workplace is categorised as a normal risk for assessing means of escape, except for the finishing area where paint spraying is conducted – this is a high risk. The assessment illustrated has been undertaken because of recent changes in the workplace which have involved changes to work processes, the introduction of some new machinery and an increase in staff from 20 to 30. The company already has a fire action plan but wishes to review their fire safety procedures in the light of the changes.

It is important to be aware that a copy of all assessments should be kept in the workplace and be available for inspection by the fire authority. It is also important that the priority action plan developed from the assessment can be shown to have been carried out.

FIRE RISK ASSESSMENT

Name and address of company: *D H Ross Ltd, Watery Lane, Toynton, Somerset*

Assessment area: *Factory unit*

Work activity: *Coach builders*

Number of people employed: *30.*

Maximum number of people in workplace: *50.*

Priority action plan resulting from the assessment

Immediate actions

- *Unblock the escape route from west fire exit*
- *Limit the quantity of fabric located in the store*
- *Remove the furniture from the upper workshop area*
- *Ensure the timber offcuts are removed from the upper workshop area daily*

Actions to be undertaken in a week

- *Reposition the exit sign in the upper office level*
- *Replace the self-closer on fire door between the office area and the workshop*
- *Identify the assembly location on the fire action notices*

Actions to be undertaken in four weeks

- *Undertake fire drills and start a record of fire drills*
- *Organise fire training and induction sessions for contractors*
- *Review the duties and responsibilities of specific employees for fire safety and update the fire action plan*

Actions to be implemented into the short term maintenance programme

- *Construct a 30 minute fire protected ceiling above the metal cutting and grinding area*
- *Relocate the flammable liquid store*
- *Remove the gas cylinder cage to a location away from the factory wall*
- *Refurbish the north fire escape stairs*

Assessor: *A N Other*
Position in the company: *Fire Safety Manager*
Date: *14 November 2000*

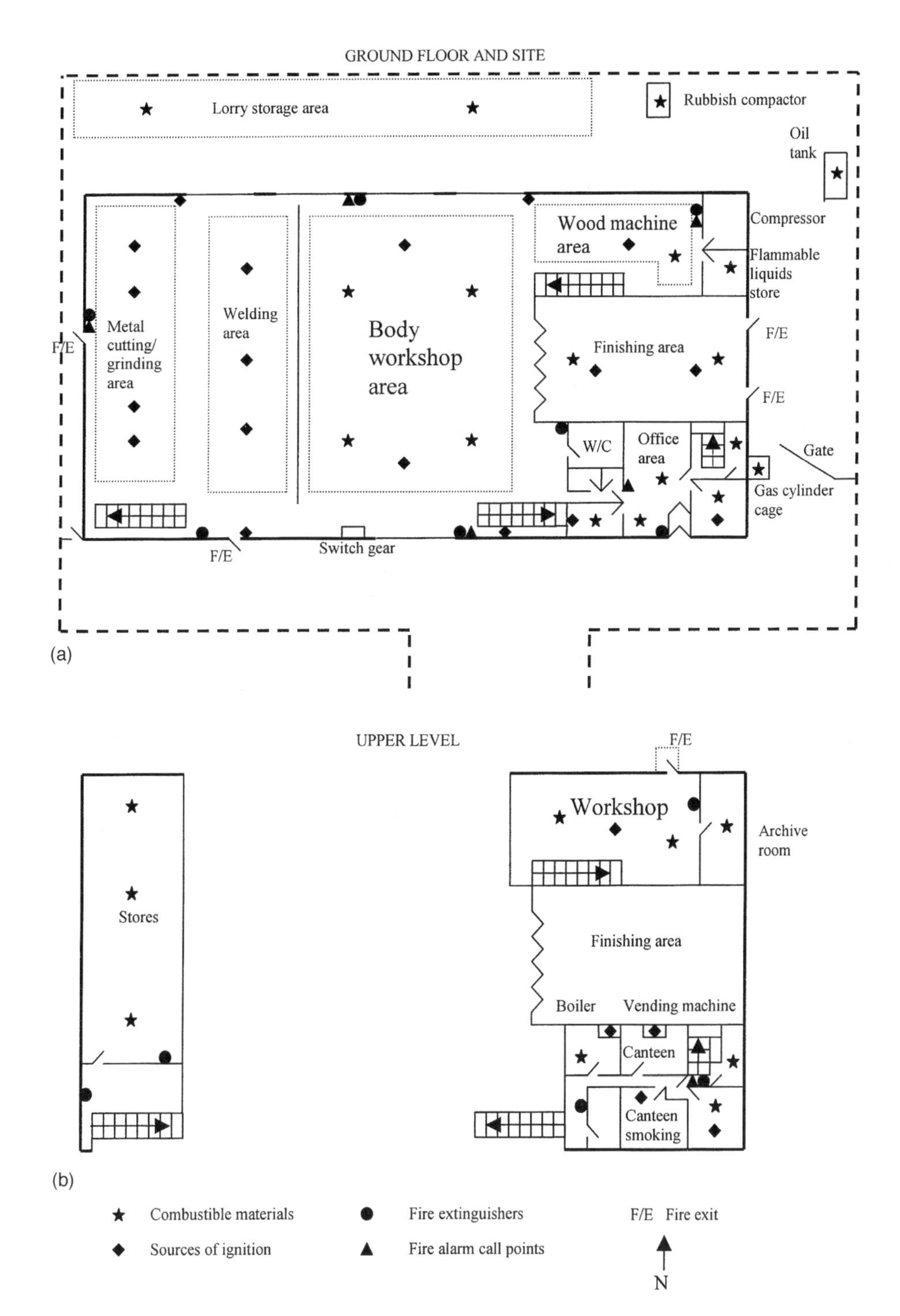

Fig. B1 Floor plans of the factory marked up during the fire risk survey.

Fire risk assessment: checklist questionnaire

Stage 1: Identify the fire hazards			
Combustible materials *(a 'no' answer highlights a concern)*			Notes:
Is the work activity free from the use of combustible materials?	~~Yes~~	No	*Timber, fabrics and upholstered furnishings used*
Is the workplace free from the accumulation of rubbish, waste paper or other materials which could be readily ignited?	Yes	~~No~~	*All rubbish and waste is removed directly to the rubbish compactor*
Are all flammable substances stored safely?	Yes	~~No~~	*In flammable liquid store*
Are the structure or fittings of the workplace free from excessive qualities of combustible materials?	Yes	~~No~~	
Is the workplace free from large amounts of stored combustible materials: timber, fabrics, furniture (containing foam padding) or furnishings?	~~Yes~~	No	*Fabrics and furniture in the north end of the store. Timber and furnishing in the upper level workshop area. Archive room*
Is the workplace free from any other combustible materials which present a significant hazard?	~~Yes~~	No	*Gas cylinders, paint, vanishes, adhesives and cleaning materials when in use*
Sources of ignition *(a 'no' answer highlights a concern)*			Notes:
Are heating appliances fixed in position at a safe distance from any combustible material and fitted with suitable guards?	Yes	~~No~~	*All radiant heaters in workshop fitted at roof height*
Is the work activity free from the processes of incineration, cooking, welding, flame cutting, frictional heat or paint spraying?	~~Yes~~	No	*Welding in designated area, metal cutting and grinding under store area, paint spraying in finishes area*
Is the workplace free from oil or gas burning equipment?	Yes	~~No~~	
Is all portable electrical equipment inspected regularly by a competent person?	Yes	~~No~~	*Register of inspections kept*
Is the wiring of the electrical equipment inspected regularly by a competent person?	Yes	~~No~~	*Register of inspections kept*
Are all flammable and combustible materials stored at a safe distance from light bulbs and fittings?	~~Yes~~	No	*Fabrics in the north end of the store close to light fittings*
Is the use of multi-point adapters and extension leads kept to a minimum?	Yes	~~No~~	
Are there suitable facilities for the disposal of smoking materials?	~~Yes~~	No	
Have measures been taken to reduce the risk of arson?	Yes	~~No~~	*CCTV cameras recently installed*
Is there a no smoking policy throughout the workplace?	~~Yes~~	No	*No smoking policy in workshops. Smoking allowed in offices*
Stage 2: Identify people at risk			
People at risk *(a 'yes' answer highlights people at risk)*			Notes:
Are there staff who work in remote areas of the workplace, or in areas of high risk?	Yes	~~No~~	*Storeman, painters in finishing area*

Cont.

Is there an area of the workplace which is used for sleeping purposes?	~~Yes~~	No	
Are there people in the workplace who are confined to bed or whose mobility is impaired?	~~Yes~~	No	
Is the workplace regularly used by people whose hearing or eyesight is impaired, or people suffering from heart ailments, pregnant women, or people with learning disabilities or mental illness?	~~Yes~~	No	
Do large numbers of people, particularly members of the public, occupy the workplace?	~~Yes~~	No	
Do people who are unaware of the fire risks and unfamiliar with the building layout and exit routes regularly use the workplace?	Yes	~~No~~	*Contractors. May be up to 20 extra people working in the workplace*
Taking account of the identified people at risk evaluate the adequacy of the means of escape *(a 'no' answer highlights a concern)*			Notes:
Are there sufficient exits of a suitable width for the number of people present?	Yes	~~No~~	*See floor plans*
Do the exits lead to a place of safety?	~~Yes~~	No	*Not the F/E on the west side*
Are all gangways and escape routes free from obstructions?	~~Yes~~	No	*Escape route from the F/E on the west side blocked*
Are all exit routes clearly lit?	Yes	~~No~~	
Are all escape route floor surfaces free from tripping and slipping hazards?	Yes	~~No~~	
Are all escape steps and stairs in good repair?	~~Yes~~	No	*Escape stairs from upper level north side starting to rust*
Are all internal doors clearly labelled?	Yes	~~No~~	
Can all fire safety and fire exit notices be clearly seen?	~~Yes~~	No	*Exit sign in upper level of the office area hidden behind a locker*
Do all exit doors open in the direction of travel?	Yes	~~No~~	
Are self-closing devices on fire doors in good working order?	~~Yes~~	No	*Self-closer on the F/D between the office area and the workshop has been removed*
Can all doors used for means of escape purposes always be open immediately without the use of a key?	Yes	~~No~~	
Is there always an adequate number of trained staff present to assist in an emergency?	Yes	~~No~~	
Stage 3: Reduce the risk where possible *(a 'yes' answer highlights options for reducing risk)*			Notes:
Can any unnecessary ignition sources (sources of heat) be removed from the workplace?	~~Yes~~	No	*Would cause considerable disruption to work processes*
Can any combustible materials present be removed, or significantly reduced?	Yes	~~No~~	*Remove some stored materials from stores and upper workshop area*
Can the general housekeeping and the arrangements for the disposal of waste and rubbish be improved?	Yes	~~No~~	*For removal of timber offcuts*

Cont.

Can additional measures be taken to prevent the occurrence of arson?	Yes	~~No~~	*Update security alarm system, improve site fencing*
Stage 4: Are the existing fire safety arrangements satisfactory for the residual risk?			
Fire safety measures *(a 'no' answer highlights a concern)*			Notes:
Are an adequate number of suitable fire extinguishers provided?	Yes	~~No~~	*See floor plans*
Are the fire extinguishers and the blankets suitably located and available for use?	Yes	~~No~~	
Does a competent person annually service the fire extinguishers?	Yes	~~No~~	*Register of servicing kept*
Where any form of fixed automatic fire suppression system is installed, is it in working order?	Yes	No	*None installed*
Is the fire alarm system in working order?	Yes	~~No~~	
Can the alarm be raised without anyone being placed at risk from fire?	Yes	~~No~~	
Are the fire alarm call points unobstructed and clearly visible?	Yes	~~No~~	
Is the fire alarm system tested weekly?	~~Yes~~	No	*Tested occasionally*
Where an automatic fire detection system is installed, is it in working order?	Yes	~~No~~	*Detection system in office area*
Where escape lighting is installed, is it in working order and maintained regularly?	Yes	~~No~~	*Escape lights installed in 1999*
Fire safety management *(a 'no' answer highlights a concern)*			Notes:
Are Fire Action notices clearly displayed throughout the workplace?	~~Yes~~	No	*Notices displayed but the assembly point not identified*
Is suitable fire safety training given to employees?	~~Yes~~	No	*Not contractors*
Are the duties and identity of employees who have specific responsibilities clearly understood?	~~Yes~~	No	*Key staff have recently left*
Are fire drills periodically conducted?	~~Yes~~	No	*None undertaken in the last year*

Fire risk assessment: summary of findings and actions

Significant fire hazards	Persons at risk	Existing measures to control hazards	Actions required
Excessive quantity of fabric stored at the north end of the store. Light fittings located close to fabrics.	*Storeman and assistant*	*Good housekeeping in store*	*Limit the quantity of fabric located in the store*
Metal grinding undertaken on the ground floor below the store. No fire protection to floor	*Storeman and assistant*	*Good working practices of employees*	*Construct a 30 minute fire protected ceiling*
Excessive quantity of timber off-cuts and furniture stored on the upper level workshop area. Potential ignition sources present	*Personnel working in the upper workshop area*	*Good working practices of employees*	*Remove the furniture from the area. Remove the timber offcuts daily*
Ignition sources close to the flammable liquid store Archive store located on top of the flammable liquid store. No fire protection to floor	*All personnel in the factory*	*Good working practices of employees. Routine inspections by the Fire Safety Manager*	*Relocate flammable liquid store*
Gas cylinder cage located beside the office area	*Personnel in the offices*	*None*	*Remove cage to a location away from the factory wall*
Highlighted deficiencies in the existing fire safety arrangements	**Persons at risk**		**Actions required**
Escape route from west fire exit blocked	*All personnel using escape route*		*Unblock escape route from west fire exit*
Fire escape stairs from the north side showing signs of rusting	*All personnel using escape route*		*Refurbish north fire escape stairs*
Exit sign in the upper office level hidden	*All personnel escaping*		*Reposition exit sign in upper office level*
Self-closer on fire door between the office area and the workshop has been removed	*All personnel escaping*		*Replace the self-closer on fire door between the office area and the workshop*
Assembly location on the fire action notices not filled out	*All personnel in the factory*		*Identify the assembly location on the fire action notices*
Fire drills not undertaken	*All personnel in the factory*		*Undertake fire drills, start a record of fire drills*
Contractors are not given any fire safety training or introduction to the factory layout	*Contractors*		*Organise training and induction sessions for contractors*
The duties and responsibilities of specific employees for fire safety are not clear as key personnel have recently left the company	*All personnel in the factory*		*Review the duties and responsibilities of specific employees for fire safety and update the fire action plan*

Index